UFO Investigation

The Methodology for a New Age

By

Richard R. Lang

Edited by Valerie S. McGregor

www.LangPublication.com

Special thanks to the Trustees of the **Lucius Farish Trust** for help and support in the publication of this book.

Date: May 2021 **Revised: Aug. 2021**
ISBN: 978-0-9883606-8-6
LANG PUBLICATION
Printed in USA

Table of Contents

Chapter 1

Introduction

Ufology is the study of **UFOs** (Unidentified Flying Objects) and especially the investigation of recorded sightings and human encounters in the current sphere of independent research.

The term **UAP** (Unidentified Aerial Phenomena) has evolved because we now understand that these objects simply do not fly in the conventional sense, although they occupy airspace above the earth, as such the phenomena has little to do with aerodynamics.

AAV (Anomalous Aerial Vehicle) is a term that Naval aviators and military aviation air controllers use as opposed to UFO or UAP.

A **USO** (Unidentified Submerged Object) is an unidentified object that is submerged underwater. The U.S. Navy and U.S. Coast guard have had their share of encounters with these types of objects. In some cases, the UAP will become a USO when the object lands on the water and then submerges.

Shapes of UFOs

The boomerang shaped UAPs in the middle on the previous page have been estimated to be over one mile across in size. Witnesses have described the shapes of UFOs varying from a sphere to a glowing tube. Some have resembled flying saucers with a lid and others triangle shapes as illustrated above. There are many reports of ball shaped orbs with reddish-orange glow, or fire-like discharges ranging in size from a soccer ball when close to the ground to much larger orbs at higher altitudes.

Incredible speed and maneuverability that is not attainable by conventional aircraft are commonly observed in most all cases.

The **Methodology** (methods, techniques, procedures) used to facilitate UFO/UAP Investigation often involves other associated anomalous phenomena to include: the anti-physical effects (observed behavior inconsistent with current physics), close proximity encounters with individuals, physiological effects on humans, psychic effects in the realm of human consciousness, and other high strangeness which will be discussed and explored in further detail later in this book.

Conventional Methodology

For at least 70 years, UFO/UAP Investigators have been scrambling to the scenes of these anomalous aerial events, with their cameras ready, donning radiation detectors, magnetometers, binoculars, and various other types of hi-tech equipment. With improving technology over the years, the lenses are faster, the cameras are far more complex, and the instrumentation is all digital, but from the standpoint of investigation methodology, not much has changed in the non-government civilian sector over the last five decades.

Authenticity in Reporting UFO Encounters

Reports of UFO encounters are often associated with very credible witnesses including Airline Pilots, Military Pilots, Astronauts, Police Officers, Fire Fighters, Military Officers, Members of Congress, and US Presidents including Jimmy Carter and Ronald Reagan. Even Thomas Jefferson wrote about a red orb that was observed in the night sky around Monticello.

There have been many sightings by U.S. and Soviet Astronauts, in 1964, three Russian astronauts aboard the Voskhod reported that they were surrounded by a "formation of fast-moving disc-shaped objects. In 1966, Gemini XII astronauts Jim Lovell and Edwin Aldrin said they saw four UFOs linked together.

During the Apollo II mission astronauts Neil Armstrong, Edwin Aldrin, and Michael Collins reported sightings of what were saucer shaped UFOs during the first flight to the lunar surface.

UFO Reports around the world

Every day ordinary people from all walks of life and in every part of the world see unusual phenomena in the airspace above this planet. In the United States alone 600 to 1000 people report such events on UFO Reporting websites every month.[1] Compound this number several times over if you include South America, Europe, Asia, and the Middle East.

Many of these events that are observed take place at great distances and high altitudes, often appearing as strange objects or lights in the night sky. Investigators estimate that about 85% of these sightings ultimately will have some rational explanation, like the unusual position of a planet, a satellite and even the International Space Station, as it could be vividly seen from the ground at night.
On the other hand, 15% of these sightings cannot be easily explained. Particularly when objects and lights that appear to be at close proximity to witnesses and in some rare cases where objects appear to be on the ground or have some physical effect on the ground.

Thousands of people have observed strange events and had close encounters with entities that appear to be non-human. They fail to report their experiences, mostly out of fear, embarrassment, or potential to be ridiculed. Even more disconcerting are witness's reports associated with close encounters that result with physiological effects like marks and burns on the body, various sicknesses that result after an encounter, and strange physic effects that occur during and after the encounters.

Witnesses typically leave the scenes of these events in shock, disoriented, sometimes sick and in utter disbelief. They often find themselves alone and isolated, afraid to tell their closest friends or even family members what they experienced.

To compound these difficulties, law enforcement officers and emergency responders are regrettably not trained to deal with the high strangeness associated with these events or the public controversy that is so often connected to these cases.

Historical References

There are countless historical references that depict ghostly bodies and supernatural occurrences that share an uncanny similarity to modern day accounts of UFOs and close encounters reports.

Ancient texts are filled with descriptions of flying machines, cave drawings and paintings in pyramids of what appear to be entities in space suits and saucer shaped space crafts can be found around the world.

The Baptism of Christ 1710

15th Century Painting

For hundreds of years mysterious objects in the sky and strange moving lights have been reported by many people including the military pilots in World War II who called them "foo fighters".

Classification of UFO Encounters

- **CE1 - Close Encounters of the First Kind**
  Visual sightings of a UFO seemingly less than 1000 feet away that show considerable detail.
- **CE2 - Close Encounters of the Second Kind**
  A UFO event in which a physical effect is alleged; this can be interference in the functioning of a vehicle or electronic device, animals reacting, a physiological effect such as paralysis or heat and discomfort in the witness or some physical trace such as impressions in the ground, scorched or otherwise affected vegetation, or a chemical trace.
- **CE3 - Close Encounters of the Third Kind**
  UFO encounters in which an entity is present, these include humanoids, robots, and humans who seem to be occupants of a UFO.
- **CE4 - Close Encounter of the Fourth Kind** is a UFO event in which a human is abducted by a UFO or its occupants.
- **CE5 - Close Encounters of the Fifth Kind**
  A Close Encounter of the Fifth Kind is a UFO event that involves direct communication between aliens and humans.

Abduction Research

Defined as Alien Abduction (sometimes also called Abduction Phenomenon, Alien Abduction Syndrome, or UFO Abduction) refers to the phenomenon of people reporting what they believe to be the experience of being abducted by Extraterrestrial Beings (ETs) and taken into a ship or spacecraft. They typically allege to have been subjected to physical examination and in some cases psychological experimentation then returned to their original environment. Individuals involved in abductions are sometimes referred to as A**bductees** or **Experiencers**.

MILABS - Military Abductions

In some cases, abductees report the presence of uniformed military officers as well as extraterrestrial beings in the spacecraft where the examinations are conducted during the abduction. Although very surprising, I've seen this in a number of reports with witness testimony to that effect.

Classified - Security Clearances

Since 1947 and the first Roswell UFO crash event a policy of absolute secrecy has been applied to protect US technological research and back engineering projects that have recovered UFO crash wreckage. This research is the most highly classified operation and the most carefully guarded secret in the U.S. Government.

Disclosure and Declassification

Anything to do with UFOs Aliens and Extraterrestrials including sightings, landings crashes, crash recovery, crash retrieval teams, and research facilities is classified way above Top Secret, only those who have the highest security clearances have access to this information.

The concept of "Disclosure" in the UFO community is that the US Government has classified and withheld all information on alien contact and Extraterrestrials. Full disclosure and declassification are being pursued by activist lobbying groups and UFO Organizations. Pressure on Government agencies to declassify information on UFOs is coming into the public purview.

Many believe that there is a point in time coming soon when the governments around the world will disclose in some form what is known about UFO/UAP Phenomena and acknowledge the existence of Extraterrestrials. More about this in the next few chapters.

Seeing Through a Glass Darkly

Metaphorically speaking, as a UFO Researcher and Investigator, I often find myself reiterating this age-old expression; "Seeing Through a Glass Darkly." These words, which originate from biblical times, are defined as having an obscure or imperfect vision of reality, as if looking into a dark mirror and not being able to clearly see the complete picture, yet somehow knowing at the end of an age the ability to do so will be unveiled.

Over 30 Years of Research

For well over 30 years, I have pursued an interest in this research, which started out from a very practical mindset, that was the consequence of flight training and a very scientific education in Aeronautics. I remember thinking that this enigma (UFO mystery) was going to be resolved with high-speed camera equipment, powerful lenses, and scientific instruments. For many years, I followed a discipline searching for physical evidence and looking for practical nuts and bolt types of answers.

Ultimately, I ended up on a journey that opened many doors and introduced me to numerous interesting and gifted people, who shared with me their bizarre experiences and told me about encounters involving very high strangeness. All this research eventually resulted in many more questions than answers and left me 30 years later with a much different understanding of physics, physical reality, and human consciousness.

Early Experience as a Pilot

The journey began for me sometime in the early part of 1976 in Florida. I was flying alone in the cockpit of a high performance, four passenger, single engine aircraft. It was night, well after dark at about 3000 feet AGL (altitude) and I was cruising at about 140 knots, heading generally north toward Jacksonville.

I was a flight student at that time, working on my flight instructor rating and an accomplished commercial pilot. I was attending Embry Riddle Aeronautical University and was to graduate with my bachelor's degree in Aeronautical Science the following year. [2]

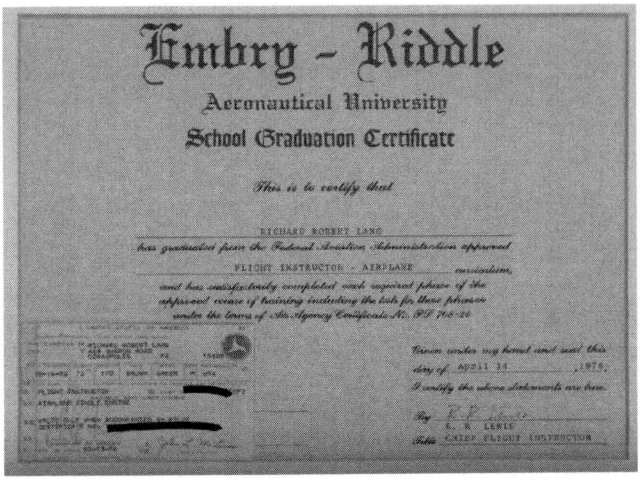

Embry - Riddle

Aeronautical University

School Graduation Certificate

This is to certify that

RICHARD ROBERT LANG

has graduated from the Federal Aviation Administration approved

FLIGHT INSTRUCTOR - AIRPLANE curriculum,

and has satisfactorily completed each required phase of the approved course of training including the tests for those phases under the terms of Air Agency Certificate No. PF 708-26

[redacted]

[redacted]

Given under my hand and seal this day of APRIL 14 ,1976.

I certify the above statements are true.

By R. R. LEWIS

Title CHIEF FLIGHT INSTRUCTOR

Flying up the Florida east coast at night is a stunning ride, particularly if you are offshore and out over the water where you have a view of the beaches and the spectacular lights from shore.

The flight that night started out much like any other adventure with a flight plan originating in Daytona Beach, a scheduled landing in Northern Florida (Jacksonville) and a round robin trip back home to Daytona. Although I had an instrument rating at that time, I was flying **VFR** (visual flight rules) that evening.

We did not have GPS back then in general aviation, so navigation was essentially facilitated with flight charts and radio navigation equipment known as VOR, which was designed to track radials or electronic lines projected from equipment on the ground.

Instrumentation in the cockpit at that time amounted to about a dozen flight instruments, including an **Artificial Horizon** (which shows the pilot the orientation of the aircraft relative to earth and helps the pilot keep aircraft in level flight),

An **Altimeter** (which is an altitude meter used to measure the height of the aircraft above the ground).

A **VOR** radio display, which is a round face instrument with a needle that centers when the aircraft is on track (aligned with the selected radial); for example, if the needle moves to the right the pilot steers to the left, in order to stay on course. (which are all pictured on the next page).

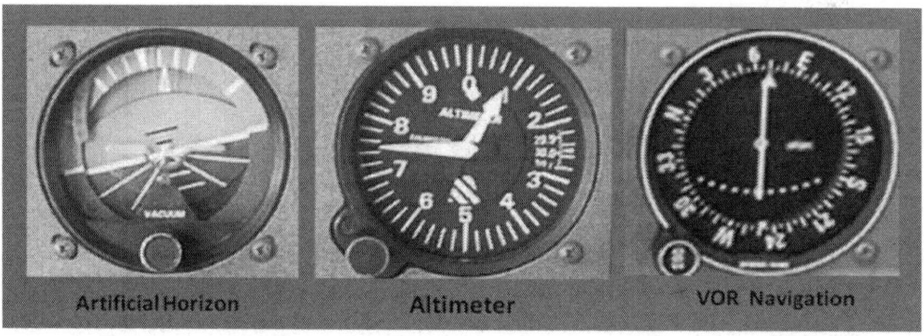

Aeronautical charts for VFR flight in 1976 were basically maps with the radials printed on them and included the corresponding frequencies that the pilot would use to tune the radio, in order to track the electronic radial. The cockpit was lighted with red light to facilitate seeing controls instruments inside the cockpit, as well as reading flight charts, without affecting the pilot's night vision.

A small aircraft of that era would typically have four radios, two **Navigation VOR radios** as described above and two **Communication Radios** for voice transmission. Some also included a **Transponder** which is a radio device which transmits a signal (code number) to identify the aircraft on a radar screen.

The GPS navigation system in my car today is probably more sophisticated by comparison to electronics in those old airplanes, nonetheless I can assure you that the pilot was understandably very busy back then with navigation, flying the airplane (keeping right side up), staying on course and talking to the **Air Traffic Controllers (ATC)**.

The voice communication radios would normally be tuned to separate frequencies, for example one radio might be tuned to the departure controller (area that you were leaving) and the other radio might be tuned to the approach controller (area of your next destination).

Aviation voice communication is very regimented and specific. There is rarely any kind of casual chit-chat between pilots and **ATC**. A radio transmission would typically start with the pilot identifying his aircraft for example, by saying EASTERN and the flight number, which would identify him as an Eastern Airlines flight.

Point being that pilots would typically be monitoring such frequencies and listening to radio conversations (radio traffic) while in flight, as I was that night. The following sequence is related from my memory of the strange event that occurred that evening:

- Out of the darkness I heard the voice of an Eastern Airline pilot make radio contact with ATC, he requested information about air traffic (other aircraft) at his flight level and in his flight path.

- After some hesitation, the ATC controller responded indicating that there was "no traffic identified on radar."

- A couple of minutes lapsed, and the pilot repeated the request, indicating they (pilots) had a visual on air traffic at their flight level. (the tone of his voice was elevated, and he seemed irritated).

- ATC responded again, awkwardly "ah. . .ah. . . negative radar contact."

- Eastern contacted ATC again for the third time, this time the pilot described the traffic (object) with words like: "huge" . . . "unconventional" . . . "hovering" . . . "stationary over the ocean" . . . "many lights on it" . . . " bright vivid multiple-colored lights" . . . "slowly rotating."

- Quiet for a moment. . . the ATC controller hesitated and nervously responded "negative radar contact" It was obvious that there was something going on, but the ATC controller was not about to acknowledge radar contact, over the radio.

- At that point, another pilot's voice came over the radio, which identified as a small business jet. The second pilot said something spontaneous like "what Eastern is seeing, then so are we!"

- The second pilot's voice sounded as if he was very irritated, when he advised that he had a visual on the same object. The second pilot did not specifically describe the object, but indicated it was the same as the Eastern pilot described. No one wanted to articulate anymore details on the radio than necessary.

After that, the radio traffic went silent! At that point, I was obviously very energized and kept looking out across the ocean trying to see it, but unfortunately it was too far out over the ocean for me to get a visual on it.

I was getting close to Jacksonville and contacted the control tower to get permission to land, which was denied! ATC (Control Tower) instructed me to maintain holding (flying in a circle) for several minutes to accommodate what appeared to be two military fighter jets that were taking off from Jacksonville Naval Air Station. I was able to observe them as they came thundering off the runway and went booming out into the night sky over the ocean, engines lit with full afterburners, in the general direction of the object the pilots reported.

The specifics of this event related herein are not as important as the three valuable lessons I learned that night:

1) There was for certain a huge unidentified, unconventional object hovering over the ocean that night, which was reported by multiple pilots, flying in two different jet aircraft that viewed it from different perspective angles.

2) The Air Traffic Controllers refused to acknowledge the presence of the object and denied that they had radar contact.

3) The Air Traffic Controllers indirectly acknowledged the object by dispatching fighter jets to intercept it.

I flew back home to Daytona Beach that night without further incident. However, this synchronistic event planted a seed deep in my consciousness and opened the door for a lifelong interest in the exploration of Unidentified Aerial Phenomena, which is still the mystery of the ages.

Purpose of this Book

The purpose of this book is to take a fresh new look at investigation technology in this New Age. This is a concerted effort to encourage investigators to look at a broader spectrum of evidence, and to initiate a framework that will more accurately categorize report databases which can be used as the groundwork for future study.

Chapter 2 - Formal Investigation Training

The book starts in the next chapter by outlining my early experience as a UFO investigator and my evolution in understanding the UFO/UAP phenomena. I am often asked how I got interested? This chapter is a general overview of my training and experience. It is a little narcissistic in the sense that it is mostly about me and how I got started in UFO/UAP investigation and research, but it serves to document my involvement and my current insight that has developed over the years about the phenomena.

Chapter 3 - Discovery Channel and MUFON STAR Team

Chapter 3 is about the original MUFON STAR Team which was a small group comprised of highly skilled investigators who were involved in filming the Discovery Channel TV Series, "UFOs over Earth" and includes a look behind the scenes while filming.

Chapter 4 - BAASS / MUFON SIP Project / STAR Team

Chapter 4 includes a general discussion about the BAASS / MUFON SIP Project and STAR Team. This was the most advanced non-government UAP/UFO investigative team in the world. I worked on the project during 2009 and 2010, which operated on both classified and unclassified levels which will be explained. This program was operated jointly out of the Pentagon and at an underground complex in Las Vegas, Nevada operated by Bigelow Aerospace, which is a defense contractor that builds modules for space stations. We now understand the Pentagon program, known as the Advanced Aviation Threat Identification Program, (AATIP) provided the funding for the BAASS / MUFON SIP Project.

Chapter 5 – Disclosure and Government Cover-Up

This chapter is an introduction to the concept of Disclosure, (when governments will acknowledge the existence and presence of extraterrestrials on earth) which is included to provide important prospective for anyone who is interested in UFO/UAP investigation.

Chapter 6 - Government Contractors, Police and Military

Chapter 6 deals with government contractors and military crash recovery retrieval teams (Recon Units) and how they operate during an accidental landing or crash. Then some advice about how to deal with Police in the field and some words of caution for when dealing with the Military Contractors.

Chapter 7 - Video in investigations

Chapter 7 is about the use of Video in investigations. The video camera is probably the most valuable piece of equipment that any investigator can have and use. This chapter provides recommendations for how and when to use the video camera while investigating cases and interviewing witnesses. A guide to different types of cameras and which ones work best for different situations is included in this chapter.

Chapter 8 Equipment

Chapter 8 is a straightforward look at the best equipment recommended for use during an investigation. This chapter covers packing for travel and having gear ready for deployment on short notice. It includes some important everyday gear to have and a look at some remarkably interesting special hi-tech equipment.

Chapter 9 - Perception of Reality

At this part in the book we are starting to push the envelope, radically thinking outside the box. Regardless of where you were born or how you were raised, you have been programmed on a subconscious level, subject to immensely powerful programming in terms of your belief systems and what you think is possible (or not possible).

Chapter 10 - Multi- Dimensional Consciousness

In this chapter we are pushing the envelope completely outside the current time space reality we live in. Ancient Teachings and recent Quantum Physics both indicate and define the existence of 12 dimensions of reality, but for the purposes of this book, we will explore the first 5 dimensions. Put your seat belt on and open your mind to enjoy chapter 10.

Chapter 11 - 6-Layer Model for Anomalous Phenomenon

Shortly after I was hired to be the Manager and Coordinator of the BAASS MUFON SIP Project, I came in possession of a whitepaper written by Dr. Jacques Vallee and Dr. Eric Davis. This 6-layer Model for Anomalous Phenomena become very fundamental in my thinking, which we will explore in detail in Chapter 11.

Chapter 12- Report writing

Probably the most important component of an investigation is writing the report. Accuracy, attention to detail, and unbiased documentation are essential to writing a high-quality objective report. Chapter 12 is a guide to writing a comprehensive report.

Chapter 13 - An SOP for Investigative Methodology

An SOP (Standard Operating Procedure) is a written plan for an investigation and deployment. Just like when Pilots use a checklist while preparing for flight, the SOP is a written plan or checklist of all the things an investigator should do during the investigation while deployed. Chapter 13 is a comprehensive SOP that can be used for the modern Investigator.

Chapter 14 - Media and the Press
I learned early on that the media and press are not your friends when conducting UFO/ UAP research. This chapter has some great insights in understanding the modern mainstream media as well as knowing how to deal with them and not letting them take advantage of you which is especially important.

Chapter 15 - Is Disclosure on the horizon
In this chapter are final thoughts I have about where we have come from and where all this is going in the not too distant future. Not to mention recent disclosure that will be revealed in June of 2021. The appropriations bill signed on December 27, 2020 mandated that a report detailing what the government knows about UFOs be released this year.

Second Book:

UFO Investigation - The Age of Disclosure
Declassification & Disclosure
What the Government Knows about UFOs
Based on 30 years of Research and Investigation
by Richard R. Lang - Published Aug 2021

This book was written for the general public, who may be interested in the UFO subject but may not know a lot about it. Unlike this first book, which was written to provide advance technical training for UFO investigators and individuals who have been involved in UFO research for many years, this book is written in general terms to permute public understanding.

Third Book:

UFO Investigation - Case Reports
A collection of short stories based on 30 years of UFO research (Expected Publication November 2021)

We've received a number of requests for books about actual UFO cases. I have laid the groundwork for a Third book which is a collection of short stories based on cases I have worked on over the years. In order to protect the privacy of the witnesses and individuals involved, some names may be redacted, and locations changed, but the substance of the occurrences and high strangeness of the encounters will be very accurately and vividly conveyed.

Confessions

One final note about the information in this book. I have made significant effort to carefully document my training, experience, associations, and certifications, in the next chapter that helped me to facilitate this research.

I have found my share of breadcrumbs along the way, where individuals knowing about certain technology, who could not directly talk about it gave me hints on where to look for information to study that helped me understand certain types of phenomena.

With that said there are some concepts in this book that I have described and noted, particularly in the technical fields involving high strangeness, that I have specific knowledge of and clearly understand. However, I will not explain how I know and understand these concepts or disclose the source of the information that was given to me.

END CHAPTER

Chapter 2

Formal Investigation Training

After graduation from Embry Riddle Aeronautical University, I worked during my professional life in the Investment and Financial sector. Throughout the late 1990s while working for a bank trust department, I was afforded time off and disposable income to explore my interest in UFO research. I was living in Virginia and was a member of MUFON. I obtained field investigator certification and started working cases. Eventually I was asked to be Assistant State Director. [3]

In an effort to develop professional investigative experience, I took some classes from the community college. I attended the state mandated training through the Virginia Department of Criminal Justice Services and obtained private investigator registration in Virginia. I maintained the registration for several years.

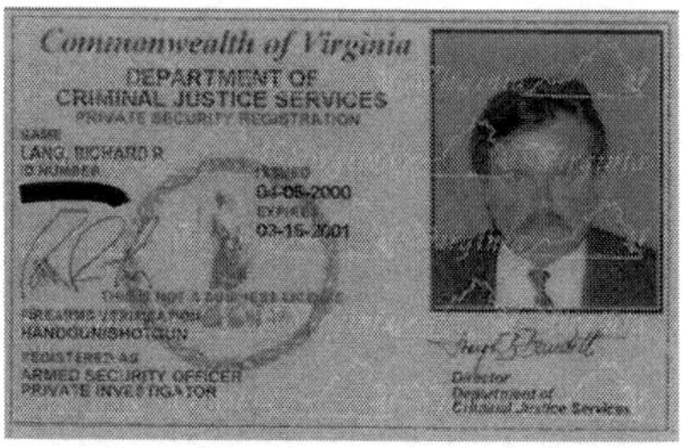

I had the good fortune to work part-time for one of the premier private investigation firms in the country located in Charlottesville, Virginia. I learned an incredible amount about investigative work, surveillance, and report writing from Samuel Yount, who is an internationally renowned private investigator, who has investigated more than 1000 cases during his 40-year career. I am forever grateful for all I learned from him. [4]

Black Helicopter

In the 1990s there was an underlying concern associated with UFO investigation. Most of us were concerned (more like paranoid) about some sort of government surveillance that seemed to be occurring. Several related incidents that occurred in my early experience are worth mentioning. One of these was documented in the "Filer's Files":

Filer's Files: MUFON Skywatch Investigations #18 (May 7, 1999)

VIRGINIA BLACK HELICOPTER - NO LIGHTS

Richard Lang - State Section Director of MUFON Virginia reports the strangest thing occurred at the lake where we live. Saturday night (5/1/99) my wife and I were returning home at 10:30 PM. We live in a gated community which has a lake in the center. As we turned on to Route 600 (3 miles from the lake) we could hear a very loud helicopter and looked over in the field to the left. I then saw two helicopters in what appeared to be a chase. These two helicopters were low in the field maybe 100 feet altitude maximum and were lighted with normal navigation lights. I saw starboard green lights as they approached. These two helicopters were chasing something that was not lighted. My wife screamed because I swerved, trying to look up, and almost wrecked the car in the ditch, running off the side of the road. These two helicopters and whatever they were chasing passed directly over us heading toward the lake. We could feel the car shake as they passed over. We drove home and could hear the helicopters over the lake.

A large heavy black helicopter flew over the street in front of our house, heading south in the clear sky. This black helicopter was at an altitude of no more than 50 feet above the treetops with no lights of any kind! It was very loud and the ground shook. As the black helicopter passed over us, it was chased by the two well lighted, helicopters we originally saw. The two smaller chase helicopters had slightly different light configurations from each other. They appeared and sounded more like medical evacuation and business helicopters. All three turned and flew over the lake again. They came back and passed over head heading south and disappeared. I turned on my scanner but was unable to pick up any information. I am a licensed commercial instrument pilot. I got a very good look at the helicopters. I was shocked they were flying so close and low in a residential area."
Thanks to richard.lang@smtpgate.crestar.com
-Editor's Note: Black Helicopters were also reported over Downers Grove, IL on March 22, at 2:00 PM at 150 feet. [5]

We were driving a new Chrysler Sebring convertible that evening, and the top was up and secured. We arrived home and pulled the car into the garage. We were both extremely nervous and shaken up because when the helicopter initially flew over our car (out on Route 600) it was so low that we thought it was going to rip the convertible top off the car, seeing that the fabric top was flopping and throbbing violently from the rotor wash of the helicopters directly overhead.

My memory of that night included another strange twist in detail that was not included in the Filer Report. We got out of the car and witnessed the helicopters from the garage (standing in the open garage doorway) as they flew over the street in front of our house.

My wife went up into the house, and I was still in the garage listening to the helicopters, when a very bright flash occurred. It was intense like a camera flash or lightning strike at close proximity. It really startled me because I could feel it (like a static snap) as well as visually seeing it. It seemed to originate from inside the garage and I instantly looked around thinking that a light bulb might have burned out and blown up (looking for a rational explanation). Then I heard my wife scream from upstairs, inside the house because she had also experienced the flash.

Strange Surveillance

We had another strange experience in that house several months later. I was investigating a case at a closed Government base in Virginia, that we know now was an NSA facility. (which I will bring to light in a moment), while my wife was home alone with our dog. Our house was located in a secured gated community around Lake Monticello (near Charlottesville). No one could get into the neighborhood without police access or an invitation from a resident.

Our yard had no grass because the builders cut down just enough trees to make room for the house. The yards were typically wooded, and mulch was placed directly around the house and small back yard. Our dog's name was Wolfe. He was a big Husky/Malamute over 90 pounds. He had a leash run line behind the house that was attached to the back deck. We would let him out of the house from a sliding glass door to the back deck and he could walk around about a 50-foot radius from the back deck, which was surrounded by thick woods.

My wife related to me later, that while we were driving to the base, our dog (Wolfe) seemed to be agitated and was pawing at the sliding glass door to the back deck. By the time she got to the door, he was extremely excited and appeared to be looking at something in the woods. She let him out by opening the door and hooking the leash run line to his collar. When she let him loose, he immediately ran to the end of his leash, pulling up, standing on his back legs and was barking at someone who was in the woods. My wife noticed a man with a walkie-talkie at some distance in the woods, who was standing there and looking directly at her, watching them.

One of the Investigators (who was a police officer) was riding with me when we returned to my home later that afternoon. When we heard my wife's story about the man in the woods, we decided to take a look around the community by driving around the lake. There was only one restaurant in the community inside the gate near the golf course. As we drove through the parking lot, we observed an unattended dark blue van in the parking lot. We actually got out and walked around the vehicle. There were windows in the back doors that were very heavily tinted but what we could see inside was typical of a surveillance vehicle. There was a sticker on the front windshield "Department of Defense - Registered Vehicle - ANDREWS AFB" which I photographed. We also observed a guest pass in the window which would indicate the vehicle was not from the community and they would have had to obtain access from the local police or with an invitation from a resident.

Later that evening we learned that several of the people involved with that investigation complained that their computers had been attacked with a virus that disabled them.

Pre-Digital Age

I am sure in that time period (1990s) some surveillance was maintained on UFO investigators in person, by essentially following them around to observe their activities and in some cases intercepting their mail from the postal service.

It should also be noted that this was an era before development of web-based Case Management Systems (CMS). UFO Investigation reports were not on the internet. For that matter, the internet was still developing, and corporations were just starting to use email for communication in the late 1990s.

UFO sighting reports were typed on paper forms and case reports were submitted in manila envelopes, that were sent through the US mail (USPS). Photographs were taken with 35mm film cameras and the film was sent to the local drug store to be developed and printed on photo paper. It should be noted that many older reports denote frustrated investigators that went to the drug store to find their film (negatives) and photo prints had been mysteriously picked up before they arrived.

The digital age was in its infancy. I remember having one of the first SONY Mavica digital cameras that used a floppy disc to record the digital images on.

Digital Age

Today information moves at the speed of light, with the internet vast information is instantaneously available to the investigator, not to mention that billions of people all over the world are walking around with the latest cellular technology and smart phones that incorporate hi-definition video cameras.

UFO case reports are posted on internet sites that are web-based, that can include digital photographs, video uploads and documents all to be included in the case report databases.

Most of the clandestine surveillance, which was done in the past by government agents, is now electronically facilitated by intercepting email, utilizing GPS features in smartphones, and through analysis of website visits, which would obviously minimize the need to physically follow anyone around.

Authors have written volumes of books on the subject of government secrecy and I will defer to them for the latest research in that area. However, we will explore a little bit more about Government, Police and Military in Chapter 5.

High Profile Abduction Cases

I was involved in several high-profile cases, one of which earned me a unique MUFON award for "Superior Investigation" at the Symposium in 1999. This was a very widely published case about a sighting and alleged abduction which took place on a now closed Government NSA base in Northern Virginia.

The following excerpt from the "Filer's Files" highlights the case: **Filer's Files: MUFON Skywatch Investigations #36 (September 9, 1999)**

MUFON OF VIRGINIA –
SIGHTING & ALLEGED ABDUCTION

VINT HILL FARMS BASE *-- Richard Lang Assistant State Director VA MUFON writes, "We recently investigated a sighting and alleged abduction which took place on a now closed military base in Northern Virginia." We initially responded to a newspaper article about the case. In corporation with the Fauquier County Sheriff's Office, we were able to locate this woman and conduct an investigation on the base. We found the following: On June 27, 1999, a 44-year-old woman was attending a camp out with her daughter. At around 2:00 AM she felt compelled to leave the camp out and told one of the other adults she was going home. Her daughter stayed all night with the other adults. She drove in her van to a now closed military base, in Fauquier County. She parked in a parking lot next to a picnic area just outside the fence. She got out of her van at 2:50 AM and sat at the edge of the parking area looking at the full moon. Then she noticed a bright blue light approaching the area. The light approached from her right side and as it got closer she noticed it was egg shaped and was the color of blue Kool-Aid. The object hovered over a clump of trees in the field about 200 yards away. The egg-shaped object was hovering on the inside of the Vent Hill Base fence. She saw* *three humanoid beings suddenly appear under the object. They were little guys, only 3 to 4 feet tall who appeared to be gathering soil samples. She remembers hearing her watch chime for 3:00 AM and the beings heard it and noticed her. They started moving toward her. Her next memory was waking up inside her van with the doors locked at 6:00 AM the next morning. She described herself as crying and shaking like crazy. She was afraid and could not remember what happened.*

Several weeks after her experiences, she told our investigators that she started to have dreams related to the experience at the base. She has remembered what she thinks happened during the missing three hours. During our videotaped interview, she said that after her watch chimed the beings began to approach her. She claims that the beings came through the fence and stood on each side of her with one behind her. The three beings took her back through the fence to the still hovering object. She was floated up in a very bright light beam that was shining down from the bottom of the hovering craft. She was led into a room and was put on an examination table. She does not know how they got her on it. They then examined her, and the next memory was waking up inside her van. We think there may be other witnesses to this event or others like it in this area and ask that they contact Richard Lang on the toll-Free Hot Line . . . Thanks to Richard Lang. [6]

Abduction Research

As fate would have it, I found myself being pulled into one abduction case after another. During those two decades, I interviewed about 30 abductees and was able to discuss with many of them, their experiences in great detail. Looking back, abduction research was uncharted territory, it was utterly fascinating and at the same time very disturbing. I vividly remember sitting in a woman's living room, hearing her talk about abduction and thinking it was so bizarre and impossible; I just wanted to get the hell out of there and go home. So much of what was conveyed seemed impossible and conflicted with my conventional understanding of physics, while as an investigator you find yourself always trying to remain objective.

Curiosity and something else that I may never be able to explain, kept drawing me into the research. It is the little similar details that seem to surface from one story to the next, that finally make you realize there is some resonant truth in the recurring details.

I continued to grow and gain knowledge while investigating UFO cases. Resounding patterns seemed to be developing through analysis of many unassociated cases, and by comparing similar events that were described by people who were not associated with each other.

Government Service - Law Enforcement

During the 1990s most of us were concerned and almost paranoid about the government. Ufology was full of stories about Men-in-Black and government Bogie-Men sneaking around behind UFO witnesses and investigators. On more than one occasion I was worried about some of the witnesses and even my own personal safety.

Fortunately, I had the opportunity to serve in several law enforcement capacities including the Federal government and came to understand better how things work behind the scenes in law enforcement and government agencies.

Sheriff's Office reserve Deputy

I was a sworn reserve law enforcement officer in Virginia with the Albemarle County Sheriff's office, which was then a reserve unit of about 20 people. This was mostly about volunteer community service on weekends and evenings, and I really enjoyed it. [7]

Official Certifications / Specialized Training

In 2001, the county sheriff provided about 10 of us with the opportunity to attend the Virginia Police Academy. Most of us were employed in secular corporate jobs and so the curriculum was structured so that we could attend night classes and several one-week courses to get fully certified as law enforcement officers.[8] After finishing the courses I graduated from the police academy on 12-21-2001 and became a **Certified Law Enforcement Officer in Virginia**, in the aftermath of the 9-11 attacks.

During my time with the Sheriff's Office, I was a member of the Search and Rescue Team and attended training with the Virginia Department of Emergency Management for **Search/Rescue Field Team Member** and then later served as a **Search/ Rescue Team Leader.**

Later On, I attended training with Homeland Security's Center for Domestic Preparedness on **Weapons of Mass Destruction Response Training** and Department of Justice classes on **Responding to Hazardous Substances**.

This was a time when Congress had mandated extremely high levels of security for USA Airports and the US Department of Homeland Security (DHS) was created.

Like many people, I felt strongly about serving my country and actually left my banking career to work in the airport at Charlottesville, serving in a dual role as a County Deputy and Special Deputy for the US Marshall service.

US Department of Homeland Security (DHS)

While working in the airport I came to know some of the senior management with TSA (Transportation Security Administration) in Virginia and was offered a job to work with the US Department of Homeland Security as a Federal Liaison for TSA stakeholders. In addition to my Airport Law Enforcement experience, I was FAA Licensed Commercial Pilot, and I had a Bachelor's Degree in Aeronautics which obviously helped me get the position.

I served from 2003 to 2005 in the airports in central Virginia. with TSA, I worked directly with the various law enforcement agencies in central Virginia including: FBI, ATF, ICE, State Police, Bomb Squads, as well as the Airport Police, City Police, County Police and Sheriffs Offices, concerning aviation security issues, federal regulation and policy for airlines and airport authorities.

During that tour of duty in a time of national crisis, I was not involved in Ufology due to the task at hand in serving my country. **Please note for the record, that my government employment had absolutely nothing to do with UFOs**. However, I learned a great deal about the inner workings of the various types of government agencies, which would later come in very handy in dealing with federal agencies and law enforcement.

Note: many of the certifications and registrations that I have mentioned in this chapter have expired and are no longer legally valid. However, they serve to support my education, experience and training in my evolution as an authority in investigation methodology.

END CHAPTER

Chapter 3

Discovery Channel and MUFON STAR Team

In 2005 after government service, I moved to North Carolina and went back to work for a bank and started researching UFOs again. I was initially Chief Investigator for MUFON of Virginia and North Carolina. Later I was appointed as State Director for MUFON of North Carolina.

The Original MUFON STAR Team started off as a small group sometime in 2007, which was comprised of a few highly skilled investigators.

In August of 2008, I was featured in a MUFON journal article as the newest member of STAR Team [9] and the following month an announcement in the Directors Message acknowledged my appointment as STAR Team Coordinator, who would facilitate communication and deployment. [10]

The original Star team consisted of 10 members. At that time, the new logo was unveiled in an article about the development of the logo.

There was some funding for expense reimbursement on a few of the cases, but for the most part it was still a predominantly voluntary operation. MUFON had a little bit of money they could use to reimburse for gasoline and mileage.

There was no formal report format other than filling out the sighting report on the MUFON website, and there was no system in place to monitor cases. When I became the Coordinator, they upgraded my access to the MUFON CMS (Client Management System) to include access to cases in all 50 states.

Most of the investigators had their own equipment. Throughout the initial investigation of the Fayetteville Abduction case (which was the first Discovery Channel TV show) we videotaped many of the original witness interviews with High-Definition video camcorders that I owned. Later Discovery Channel actually used some of that video footage in the TV productions.

Pictured here I am with my good friend Jeremy Ray who was a Star Team investigator and later featured in many Hanger One Episodes for TV.

As far as any protocol for working on cases, there really wasn't any. The cases came up somewhat randomly, typically someone would send a heads up (email) about a special case to James Carrion and he would ask me to look into it and send someone out to work it.

The Discovery Channel shows were the first real deployments involving the STAR Team. The show producers paid for lodging and travel expenses for some of the investigators during the filming.

James Carrion was the international director of MUFON. We are pictured here, together with the late Budd Hopkins during the MUFON Symposium in Denver.

One evening during the Symposium I had the good fortune to spend an evening with Budd Hopkins. We shared a table at dinner in the hotel and spent several hours together.

He told me all about how he got started and all about his incredible work with UFO Abductees and regressive hypnosis techniques he developed.

I will always remember that evening I spent with him. He was an incredibly brilliant person who did some extraordinary research.

It was James who initially appointed me to be one of the original star team members and then later appointed me to be the coordinator of the STAR Team to manage the operation.

We are pictured here while filming in Fayetteville North Carolina. I worked closely with James and had great respect for him then and I still do for that matter. James was an extremely intelligent individual who owned and operated an IT type of business and I knew he had honorably served his country in the US military as an intelligence officer. We worked together on the cases which were featured on the Discovery Channel.

Discovery Channel TV Series - UFOs over Earth

At that time, a TV production company was working with MUFON and sought to produce some UFO related shows for the Discovery Channel. We sent them some of the video and copies of reports that were written about the cases and they decided to do a TV show called "the Fayetteville Incident". When the filming started, I worked with the producer's crew to coordinate filming in Fayetteville.

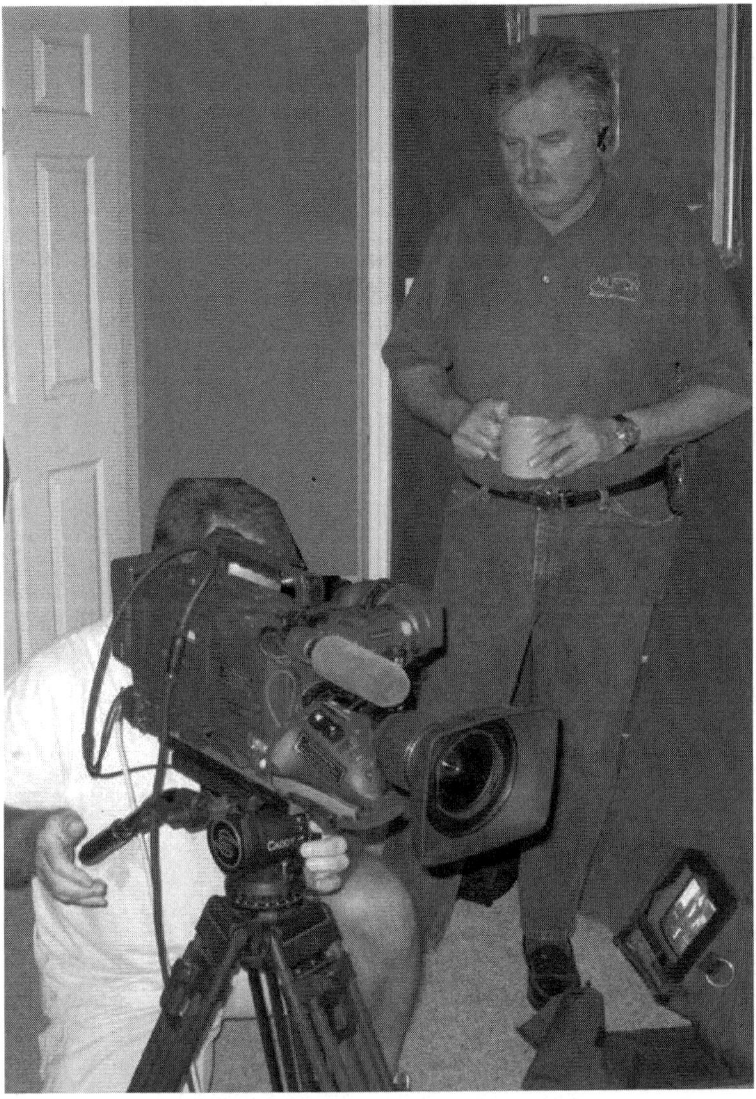

Pictured here, I am watching the production crew camera man as he is videotaping an interview with a witness.

I spent most of the rest of the summer on weekends and utilized a couple of weeks' vacation time to be on location with the crew while filming the show.

First Discovery Channel Show - Fayetteville Incident

In 2008, We investigated the case in Fayetteville, North Carolina which became world famous. The initial case report came in and was investigated locally by a field investigator who wrote an interesting and compelling case report, which essentially laid out a story that involved five witnesses who were fishing on the bank of the Cape Fear River and a close encounter which allegedly led to an abduction experience.

The primary witness Chris Bledsoe was the subject of the initial report. I traveled to Fayetteville and interviewed Bledsoe on video tape. He is an intelligent man and has a very technical mind. We were able to relate to each other because we are both pilots. He had given me a lot of insight about the abduction phenomenon. Fortunately, he was able to help us locate the other witnesses.

We were able to locate and interview all the witnesses and video tape their testimony. We found each witness, one at a time and interviewed them separately. Each time we videotaped their story and compared it to the testimony of the original witness. The testimony of the witnesses collectively was fairly consistent and corroborated Bledsoe's story. The more information we were able to collect, the more interesting the case became.

Second Discovery Channel Show - Bucks County Flap

Later that year a second episode in the series was filmed in Bucks County Pennsylvania, and I was asked to participate in it with the State Director who led the investigations.

Bucks County was suddenly flooded with UFO Sighting reports. Radar was obtained from the Philadelphia International Air Traffic Control Tower on a huge six-sided craft with no transponder and trace evidence was discovered in the yard of a woman who claimed a triangular craft dropped probes into her yard. The entire **"Fayetteville Incident"** [11] and the **"Bucks County Flap"** [12] shows are available for viewing on YouTube. See references for links to video of the shows.

Filming behind the scenes with Discovery Channel

It is interesting how the producers film a documentary. Typically, you will see a witness talking in a scene with a black background, where only the head and shoulders of the individual are shown on the screen, like a talking head scenario. When they film something like this, it is done in what they call a black drop.

Pictured on the next page, I am being filmed in the black drop setting. Behind me is a soft black curtain, above my head is a shotgun directional microphone, in front of me is the camera and the producer. On the floor under the camera is a small monitor that shows the producer what was actually being filmed.

The producer sits next to the camera and asks the investigator questions.

When they edit the video, they cut out all the producer's questions and all you end up seeing and hearing is when the investigator describes something which was usually an answer to one of his questions.

On the next page, this second picture shows me sitting in the same black drop setting in front of the camera.

There was a large television monitor on the floor next to me (bottom right) which shows what was being recorded and exactly what it will look like when viewed on TV.

I estimated that I spent about $1,500 cash out of pocket for gasoline and other expenses associated with the investigation and filming, but what I learned about the media producers and how film documentaries are produced was priceless.

No Longer Associated with MUFON

Over the last few years, I have received many phone calls and email requests, asking me to discuss details about the BAASS / MUFON SIP Project. For the record, I currently do not have any official association with MUFON organization. I am no longer a member of MUFON and not an employee.

END CHAPTER

Chapter 4

The BAASS / MUFON SIP Project

In 2008 I published a personal blog titled ***"from the Dark Side of the Looking Glass"***. I have always thought this whole UFO/UAP research thing is kind of like when Alice went through the looking glass in Wonderland. There is a rational normal life outside and a dreamlike altered state of reality on the inside of the mirror, and so it was with BAASS and the MUFON SIP project.

BAASS / MUFON SIP Project

During the period from February 2009 until February 2010, it was my privilege to serve as the Manager of the BAASS / MUFON SIP Project and STAR Team. As indicated, this was the most advanced non-government rapid response UAP/UFO investigative team in the world.

James Carrion appointed me to the position, and I served on several planning group meetings in Colorado prior to the start of the project with him.

Behind the scenes the Defense Intelligence Agency (DIA) was secretly preparing to fund several UFO/UAP research programs which were ultimately associated with MUFON through the Pentagon's **Advanced Aerospace Threat**

Identification Program (AATIP), which we did not know about at the time.

The ATTIP program operated on both classified and unclassified levels. It was operated jointly out of the Pentagon and at least for a time at the underground complex in Las Vegas, Nevada managed and owned by Bigelow Aerospace.

In 2009 Bigelow Aerospace was an American space technology company, based in North Las Vegas, that manufactured and developed expandable space station modules and is owned by Robert Bigelow. [13]

BAASS (Bigelow Aerospace Advanced Space Studies), was a sister company to Bigelow Aerospace Corporation and a research organization that focused on the identification, evaluation, and acquisition of novel and emerging future technologies worldwide, specifically related to spacecraft. BAASS was headquartered in the secure underground complex at Bigelow Aerospace in Las Vegas.

In January of 2008, BAASS was registered as an LLC in Nevada. BAASS was created in part to secure the bid for the research project and get funds from AATIP.

At that time **MUFON** was the largest international UFO research organization in the world. Over a period of thirty plus years MUFON had created a very complex data base comprised of UFO witness reports and investigations. These reports were submitted by witnesses from the general public and accumulated on a web-based MUFON site known as CMS (Case Management System).

In February 2009, an agreement was finalized between BAASS and the MUFON Board, which created a program funded by BAASS, where MUFON was subcontracted to provide case report information from MUFON's CMS website directly to BAASS.

Per the terms of the contract agreement MUFON provided data from sighting reports that were submitted to MUFON CMS website by the general public, in exchange for funds paid directly to MUFON each month from BAASS. This Contract was sometimes referred to as a Purchase Agreement (purchase of information) and the project was referred to as the SIP Project (STAR Team Impact Project).

This was the unclassified part of AATIP. At that time, we did not know about AATIP or where the funding was actually coming from. We were aware (at least I knew) that the funding did not come from Robert Bigelow personally but was provided by some other agency which was not disclosed to me.

MY Side of the MIRROR

A good place to start might be to understand how the SIP Project worked on my side of the mirror, which was not classified.

BAASS-MUFON SIP Project

During that year, while working on the SIP Project with BAASS, I learned an incredible amount about how to investigate and study this Phenomena and I developed relationships with some of the most experienced UFO Investigators, Researchers and Ufologist in the world.

The purpose of the BAASS / MUFON SIP Project was in part to provide funding to professionally investigate UFO/UAP events and make available information from the MUFON witness sighting report database (CMS) directly to BAASS.

Part of this funding was also allocated to payroll a team of full time Dispatch Operators, who would work continuous shifts throughout the week and monitor the incoming sighting reports.

Shortly after being hired to manage the SIP project, I came in possession of a whitepaper that was written by Dr. J.F.

Vallée and Dr. E.W. Davis, that is titled:
Incommensurability, Orthodoxy and the Physics of High Strangeness: A 6-layer Model for Anomalous Phenomena". [23] which will be discussed in more detail in chapter 11.

I don't know if this was synchronicity or if John Schuessler really had the vision and foresight to know where this was headed for me, but in any event, he was the one who gave me the white paper.

John Schuessler was an aerospace engineer with NASA and one of the original founding members of MUFON who served on the MUFON Board for many years. John was always supportive of my endeavors and I will always consider him my friend.

The "Catbird Seat"

During the early stages of the SIP project, I studied this white paper intensely and realized that it was totally right on the mark as to investigation and research on UAP phenomenon. I used the concepts in the paper as a framework for my thinking while performing in my role. Keep in mind we were receiving 450-850 case reports per month and we were literally looking at this information night and day (7 days a week) for about a year.

Metaphorically speaking this was the "Catbird Seat" if there ever was such a thing in the Ufology Community. Throughout that year, I took advantage of every minute, to study this phenomenon and learn all I could about how to investigate sighting reports and how that data might be effectively used.

February 2009 - SIP Project start-up

As the SIP Project contract was finalized and approved by the MUFON Board we were in the final planning and development, which included a PowerPoint presentation which I prepared for the State Directors and the membership that was introduced to outline the SIP Project and invite member participation.

Project to run like a Business

At that time MUFON was an organization of volunteers with a wide variety of members, with varying degrees of interests that were focused in many different directions. MUFON was also very decentralized through a system of State Directors and for the most part supported by its membership and benefactors.

The idea of using the STAR Team to separate the funded BAASS Project business from the general membership was taking shape. Since we suddenly had the capacity to pay people, we wanted to control the activities of the project and run it like a business, which made sense from an operational standpoint. So, it was conveyed to the membership that applications were being accepted for prospective employee positions for Investigators and Dispatch operators.

Press Release

"***MUFON has now taken UFO Research to its next level This is possible through a funding arrangement between***

MUFON and BAASS (Bigelow Aerospace Advanced Space Studies). MUFON is the largest UFO research organization in the world and is the logical choice to carry out this mission."

By mid-April we were definitely in the game and the SIP project was up and running. Like most things in life, money means everything and the funding from BAASS really did take the STAR Team up to the next level, way up! For the first time in history Non-government UFO investigative teams were going to be paid to conduct research.

Dispatch Operators

The SIP Project provided money to pay for part-time dispatch operators. The dispatchers were paid hourly ($15.00) and worked continuous shifts 24 hours a day 7 days a week, monitoring sighting reports as they were coming into the Case Management System (CMS) and then were vetting the cases.

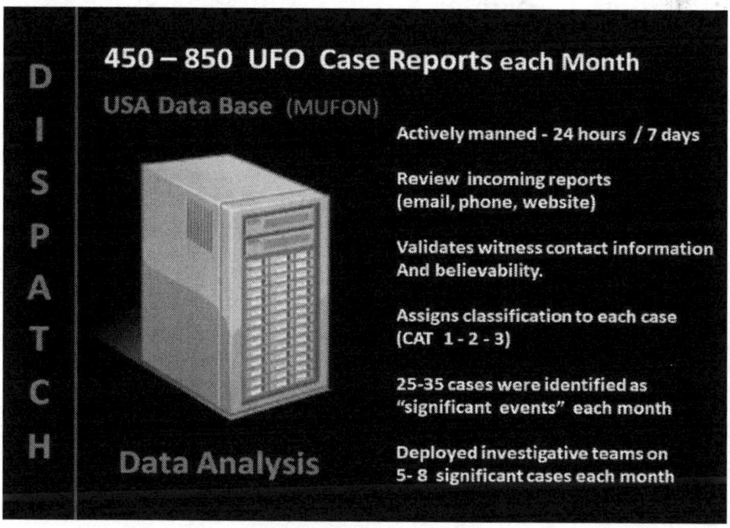

Every sighting report was categorized based on a set of criteria and protocols that was established at the onset of the program. These protocols were fine-tuned after several months with the help of our friends at BAASS.

Dispatchers verified the witness contact information on all incoming cases, which was a first for MUFON. They would actually contact a new witness and validate the witness contact information to establish believability. In many cases, they would have discussions with the witness to clarify the details of the narrative. A lot of very specialized training was done with the dispatchers, on an ongoing basis.

We encountered our share of scammers and pranksters using these sights to submit bogus reports.

Verifying Credibility of Witnesses

There are some very useful internet-based white-pages type of anti-fraud websites that verify names, addresses and phone numbers. Simple checks running phone numbers through anti-fraud websites can save a lot of time and energy and at the same time help to verify credibility of a witness.

I remember one evening a very interesting report came in that was allegedly submitted by a Catholic Priest. The dispatchers on duty ran a check on the phone number that was submitted with the report. Information came back including an address, which on google maps was in the middle of a large Catholic church complex. That is the kind of thing you want to see when trying to verify the sighting report.

Case reports were vetted by the Dispatch team based on criteria that was provided by BAASS as follows.

Dispatchers assigned a Vallee classification (CAT 1-2-3) to each case by using the following Questions:

1. Did you witness a strange or unidentified light(s) or object(s) in the sky?
2. Did the light or object land or affect in some way the immediate physical environment?
3. Did any of the witnesses' experience one or more of the following during or after the event?
    - A) Observations of beings or entities
    - B) interaction with nonhuman entities
    - C) episodes of missing time
    - D) wounds or trauma to body of unknown origin
    - E) taken onboard a craft
    - F) telepathic communication
    - G) out of body experience
    - H) encounter with an unknown animal

CAT 1 = Question 1-Yes Question 2-No Question 3- No
CAT 2 = Question 1-Yes Question 2- Yes Question 3- No
CAT 3 = Question 3 -Yes

Significant Reportable Events

Dispatchers also identified cases that were considered significant based on the following criteria. Cases that qualified were sent to BAASS as Significant Reportable Events. A list of about 25-35 Significant Reportable Events were reported and sent each week. From the list of Significant Reportable Events, I would select about 5-8 cases to deploy investigators to work on the cases.

Note that the criteria for Day and Night are different.

Daytime Observation

A) Multiple Witness sighting below 600' AGL (above ground level)
B) Any sighting within 600' of a witness
C) Any landing event, (must see it hit ground)
D) Any event with ground trace evidence (photos/video/recordings do not count as trace evidence)
E) Any sighting below 600' AGL with photo/video evidence
F) Any sighting reported by police officer, military service member, professional pilot, air traffic controller Astronomer (other professionally trained observer whether on or off duty)
G) Any sighting of object(s) with associated electromagnetic effects or (car stops, clocks stop, electric appliances failure) other associated anomalous events. (entity inside)

Night Observation

A) Multiple Witnesses within 600' of the object(s)
B) Any landing event (must see it hit ground)
C) Any event with ground trace evidence

(photos/video/recordings do not count as trace evidence)

D) Any sighting reported by police officer, military service member, professional pilot, air traffic controller. Astronomer or any other professionally trained observer (whether they are on or off duty)

E) Any sighting of object(s) with associated electromagnetic effects

Looking at the whole Northern Hemisphere

One of the most significant aspects of the Dispatch Operation was the centralized effort being able to look at sighting report data, coming in from all over the United States, through a single point of reference.

We did not completely realize at first, but as time passed, focusing on significant aspects of cases and the culmination of similar occurrences across the whole country, we were seeing some interesting data starting to materialize. which yielded very useful information.

We learned how powerful CMS is in terms of the scope of sighting reports and reliability of information, for example:

On a Sunday Morning in October 2009 MUFON Dispatch received approximately 70 sighting reports (during the overnight) describing a similar object and event. Witness descriptions were similar, and reports came in from Virginia to Nova Scotia:

- Witnesses generally reported a Bright Ball like light with a Cone shaped light coming from it, shining down to earth (also referred to as flood light and flashlight beam)

- Generally observed for about 30 seconds
- Sighting reports timed the event from 1945 hours to 2010 hours EST
- Object reported to be high in the sky, general area where jets normally fly
- Some Witnesses reported that they could see a triangular shaped object in front of the light coming down to earth

Note: There was a scheduled Rocket launch at the NASA Goddard Space Flight Center`s Wallops Flight Facility, located on Virginia`s Eastern Shore. The time of the launch was consistent with the times of the multiple sighting reports coming into CMS.

Associated Press Article - NASA launches rocket, dozens report strange lights. (AP) 12 hours ago WALLOPS ISLAND, Va. NASA says it successfully launched a rocket in Virginia as part of an experiment, and the blast may have caused dozens of people to report seeing strange lights in the sky.

CMS was an excellent and immensely powerful reference source that records what the public is seeing (or not seeing) in the sky over the United States.

Operations Manager

Another first was the establishment of a paid operations manager position (full time employee). The operations manager made all arrangements for airline travel, car rental, hotel rooms, meals and equipment. Everything was paid for up front. Each investigator was issued a VISA money card, so they incurred no out of pocket expenses.

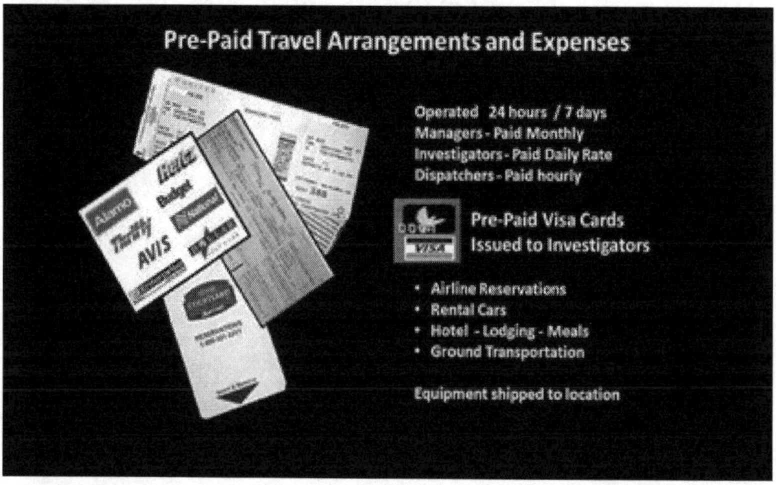

During the deployment, the operations manager would make airline reservations, book hotel rooms and deposit money on the investigator's VISA card to cover all the expenses. When a deployment order was issued, the equipment needed would be referenced, so operations manager could make arrangements to send it to an airport or hotel.

Investigators

Investigators were paid $100.00 per day for their time and $40.00 a day per diem for meals, plus the cost of all airline travel, rental cars, and hotel accommodations. The local State Coordinators were also paid $50.00 for each case to help coordinate the investigation on the local level.

Web-based Schedule

A scheduling system, which was all virtual (web-based), containing information about each investigator on stand-by, specifically listing their availability times, areas they were able to travel, and it was organized by geographic location (you do not want to fly someone from California to work in Virginia). The dispatchers were also on the scheduling system.

Equipment

We purchased some high-tech equipment that was packed in waterproof shipping cases for immediate delivery to the location. Typically, equipment would be shipped to the hotel or delivered to the airport baggage area, waiting for the investigator to arrive. More about equipment in Chapter 8.

Formal Job Descriptions Job descriptions with specific requirements for each position were established and as seen below:

SIP Coordinator and Manager (My job)

Job Duties:

- Supervises Dispatcher Operation.
- Deploys SIP investigators to the scene of a MUFON Category 2/3 cases.
- Overseas Deployment and Reviews Case report from Deployment.
- Coordinates with BAASS personnel about deployment schedules and other operational requirements.
- Communicates Significant Reportable events to BAASS.
- Processes daily status reports into weekly and monthly status reports to be sent to the MUFON International Director and BAASS.

- Writes monthly STAR Team Report for MUFON Journal outlining Significant Cases and recognizing Investigators and State Coordinators.

SIP Operations Manager

Job Duties:

- Arranges all travel (airfare, hotel, car, per diem, etc.) for designated SIP personnel.
- Keeps detailed accounting of all SIP expenditures and submits weekly and monthly.
- Reports to the MUFON International Director.
- Handles all payroll for SIP salaried employees and independent contractors.
- Handles all equipment shipping for a SIP team deployment.
- Other administrative duties as required.

Dispatch Operator

Job Duties:

- Reviews incoming reports (email, phone, CMS)
- Validates witness contact information and believability.
- Assigns an initial Vallee classification to each case (CAT 1-2-3).
- Identifies Significant Reportable Events that will be sent to BAASS.
- Updates CMS with DISPATCH NOTES.
- Communicates with Sip Manager about Category 2/3 cases and Significant Reportable events.

- Dispatchers are NOT to investigate cases! (pave way for Investigator to investigate case).

Identification ID Badges

STAR Team investigators received special customized engraved identification ID Badges to go with the MUFON ID Card, which were required to be used while on deployment.

Deployment operations

The SIP Project was operational for over 11 months. During that time, we deployed paid investigators on 65 cases. On more than one occasion two or more investigators were deployed on a given case.

We also sent data in the form of Significant Reportable Events (35-40 per month) to BAASS as well. At last count, we had about 80 STAR Team Contract Investigators and 8 Dispatchers.

Significant cases that fit very carefully developed criteria for the project mentioned above were vetted and reviewed for possible deployment of a rapid response investigative team.

In the context of this book "Deployment" is defined as the operational process of putting an investigator or team of investigators at the scene of an event. This includes the process of selection (who is going to work on the case), the SOP (standard operating procedure) for the deployment, arranging transportation, lodging, making equipment available, organizing support personnel and the report process. Deployment is expensive and requires funding to support an operations staff in order to be facilitated effectively.

Part of my responsibility was to make decisions about deployments for the SIP Project. Deployment was solely my responsibility and I had complete control of who went where, when and for how long. No one at BAASS ever interfered with us in this regard!

Throughout a deployment, things happen very quickly, and you must be organized, well prepared in advance and have an SOP in effect. There is relationship between the time that elapses, from occurrence of an event to the time the investigator is on location, and the quality of information that can be obtained and developed. The sooner an investigator gets to the scene the more accurate the report is generally going to be.

I remember a case that came up in Texas. We deployed a local investigator to work on the case. I was in North Carolina at around 10am when the decision for the deployment was made. I decided to work on the case with the investigator and deployed to Texas from North Carolina. By 7pm I was standing in the witnesses back yard with the other investigator, with a case full of equipment.

Confidentiality

The MUFON / BAASS SIP project investigators all signed a confidentiality agreement which stipulated while the case was being investigated the elements of the case were to be kept confidential. (copied on the next page)

We did not want a lot of internet traffic about a case (or our witnesses) going on while we were interviewing and collecting evidence. This is especially critical when dealing with anonymous witnesses or law enforcement personnel who might be willing to help you, but do not want to be in the public view. This is an example in the MUFON CMS system for a deployment requesting confidential handling of the case posted by me.

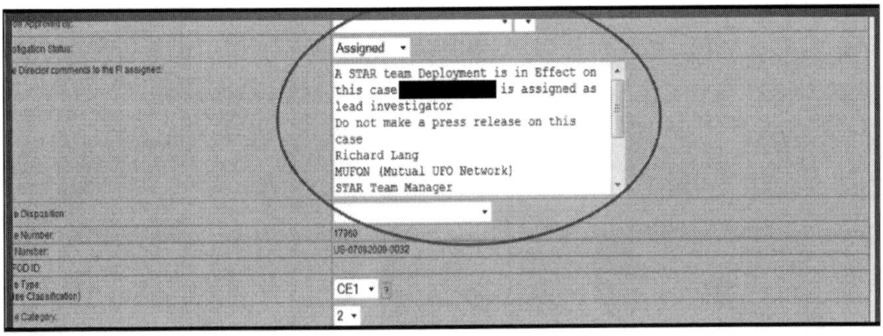

SIP Confidentially / Non Disclosure Agreement.

The following conditions of confidentiality must be agreed to by all applicants to be considered for STAR Team Deployment under the SIP Project. STAR Team Deployments are strictly confidential and must be maintained at the highest level of security.

- Strict confidentiality is in force from the moment of deployment notification. Discussion of destination and travel method in communications within your immediate family is acceptable; however friends or other associates are not to be informed of any SIP Team deployments.
- During a SIP deployment, any and all communication related to the deployment (case) should be directed only to the, the SIP Coordinator/Manager, the State Coordinator, the STAR Team Chief Technical Advisor, the MUFON International Director or the Director of Investigations.
- There should be no media communication of any type while on a SIP deployment, including newspaper reporters, television, radio, email websites or other sources or places of public discussion. Investigators and SIP Staff should refer all media inquiries to the MUFON Director of Public Education.
- All information and data including but not limited to the following: notes, reports, witness writings, measurements, photographs, video, drawings, renditions, witness testimony, etc. that are assimilated during the SIP Deployment, are the sole property of MUFON International. (see note below)
- Information and Data that is discovered during an SIP Deployment should not be used in writing books or other publications of any kind, with out the express written permission of the international Director. (see note below)
- Media Release Forms must be signed in all cases where material and evidence is obtained from witnesses or other individuals.
- During the course of a deployment any information, evidence, photograph or other data that is assimilated should not be surrendered to any Law Enforcement or Government Agency without approval from the MUFON International Director or by Due Process of Law. (MUFON personnel are expected to obey all Federal, State and Local laws) This provision is not meant to prevent SIP Investigators from corroborating with law enforcement when in their discretion it is necessary to properly investigate the case.
- Photographs, video and other data provided by a witness shall remain the witness property unless surrendered to MUFON by written agreement.

Note: All data collected during the SIP deployment is confidential. Any information and data form a SIP case that is published in the MUFON Journal or is used in the Symposium proceedings can be considered available publicly and is released from the restrictions in agreement.

I agree to comply with the conditions of confidentiality outlined in this document and I certify that the information contained in my SIP Application is true and accurate.

Written Name: ________________ *MUFON ID #* _________ *STATE:* ________

Position: SIP FI ( ) SIP State Coordinator ( ) SIP Dispatch Operator ( )

Signature: ______________________________________

Date: March 20, 2009

This is the exact agreement that all the Star team members signed while I was managing the program.

Information Released and Available Publicly

Most important is the note at the bottom of the agreement pictured on the previous page. Note: All data collected during the SIP deployment is confidential. Any information and data from a SIP case that is published in the MUFON Journal or is used in the Symposium proceedings can be considered available publicly and is released from the restrictions in agreement.

When the case investigation was complete and closed it was agreed that the information would be published in the MUFON Journal and hence be public information for research. Each month I sent a summary of all the cases the Star Team deployed on that were published in the MUFON journal.

Breach of Confidentially can compromise a great case.

An event occurred in Louisville, Kentucky on the premises of Louisville International Airport. The following excerpt will provide some insight. MUFON Dispatch verified a report submitted by a ramp (cargo) employee at Louisville International Airport. The event occurred during daylight while witness was on duty at the airport. The witness heard ground radio traffic talking about a UFO, and about a 1/2 hour later witness sighted an object and described the following:

- A reflection as if someone had a mirror and was trying to signal you.
- It would flash then be dark and then the sun would catch it and flash again.

- It was about midway over the runway at this point, estimated altitude 600-1500 AGL.
- As object came closer, witness could make out it was rotating and was a cylinder and then a diamond shape. As it spun horizontally and clockwise the cylinder shape would catch the sun and reflect, but the diamond shape was black.
- The speed of the rotation was about 4 times a minute.
- Numerous airport personal sighted object including, several Airport Public Safety Officers (Emergency Police / Fire dual role Responders) who videotaped the object in flight over airport.

One might be thinking "It just doesn't get much better that this!" with multiple highly qualified observers at an airport including police, emergency responders and even video tape . . . **well not exactly**, **what happened?**

Behind the scenes (almost immediately) someone at MUFON filed an official FOIA request with the FAA. A few days later, excited investigators who had contacted sources at the airport and made appointments to conduct interviews, were stopped dead in their tracks, as might be expected, suddenly no one would talk about it, like it never happened!

Lesson learned; Move fast, deploy quickly, and maintain confidentially while the investigation is in progress. "you can get more done in the shadows, than you can in the spotlight".

Confidentiality agreements changed after SIP Project?

I believe the confidentiality agreements for the Star Team Investigators may have changed after I left, and the SIP Project ended. Based on a blog article written by the late Elaine Douglass in May of 2011, ***"The Gagged-for-life Star Team Confidentiality Agreement"***. It appears that much stricter confidentiality agreements were put in place. You can read her blog article. The link can be found in the references at the end of the book [14]

SOP (standard operating procedure)

When we first started the SIP project there was so much excitement in the air, and everyone was ready and anxious to deploy. Keep in mind even the most experienced airline pilots still use a checklist and a good SOP with a set procedure for how a deployment should unfold is an absolute necessity. In Chapter 13, we will look at more about SOPs.

On the BAASS side of the Mirror

During this same time, Bigelow had basically created his own UFO/UAP investigative team at BAASS. They hired "50 top-notch scientists and experienced law enforcement investigators to do extensive lab analysis as well to conduct field investigations. Behind the scenes they had a 22-million-dollar budget and operated over a three-year period.

I am sure that a lot of the work done on that side of the mirror was classified. Recent disclosures by the Defense

Intelligence Agency, have confirmed that they were testing wreckage recovered from UFO crashes, at Bigelow Aerospace in Las Vegas, which would have obviously been highly classified. We did not know anything about that back then. (more about this in Chapter 15)

BAASS ordered 40 MUFON Field Investigator manuals ($1,800) for their team. The manuals were used to educate the BAASS investigators on the UFO topic and familiarize them with the methods of their MUFON partners.

Propulsion was the Prime Objective at BAASS

The focus of the project was to obtain information and evidence about propulsion, so events that involved objects in close proximity to buildings or the ground were of high interest.

Of particular interest were case reports that involved electronic equipment malfunction, like automatic lighting systems being activated or security alarms going off, during an event in progress.

For example, a report that involved an object hovering over a building which caused the outside HVAC units to malfunction or burn out was the kind of thing we were looking for.

The magnetic signature on a vehicle or a cell phone battery that mysteriously discharged during an event would be very valuable. Also, any evidence having to do with any kind of human injury like burn marks was carefully analyzed.

Photographs would be of high interest if the photo or video depicted halos of a specific color. Halos on photographs can be used to hypothesize the possibility of ionized fields or electromagnetism.

No Interest in Abduction or Cattle Mutilation
Cases about abduction or cattle mutilation were of little interest to the BAASS team and were not typically submitted as Significant Reportable Events unless some element of the case involved evidence of Propulsion. I was still free to deploy investigators and pay them to work on these kinds of cases, but the reports were not sent to BAASS.

The BAASS team conducted their own investigations after we submitted our SIP investigation reports. After we deployed SIP investigators on a case, upon completion of their investigation, the investigators submitted a written report to me with witness statements, investigator narrative, photos, video, notes, dates, times, locations, and GPS coordinates. I would then prepare a very detailed report based on a stand format which was submitted to BAASS.

We often video-taped witness interviews and uploaded to what we now call the cloud, so the BAASS team could look at the interviews first-hand and make their assessments. (more about videotaping witnesses in Chapter 7)

The BAASS team investigated cases privately and very often deployed investigators to re-evaluate information, interview additional witnesses, and conduct supplementary scientific and medical tests.

Shortly after the start of the SIP project, MUFON CMS was modified to include the **"third party Release Agreement Question"** which came up during the initial internet submission of every sighting report. This question allows the witness to decide if they want to have their personal contact information released to third parties like BAASS. (Note: you could see this question directly under the witness narrative section).

Those individuals (witnesses) who chose to answer "YES" to release of their contact information to third parties, essentially agreed to having their contact information shared with third parties and would be subject to being contacted by such third parties as BAASS. Note: witnesses answered YES in about 95% of case reports.

BAASS was given full access (password and user ID) to the CMS data base, in all the cases where the witness answered "YES" to the "third party release question" their investigators could see all that case information in CMS including witness contact information. BAASS access to CMS was to view information only, they could not alter or add anything.

BAASS did not share findings or test results with us.
They were paying us generously for information, while providing enough funding that we could freely investigate any cases we chose to and write reports about anything we desired without restriction. However, the deal clearly stated that BAASS was a private organization, and their research was proprietary information. Not to mention that some of the

work they were involved in was highly classified. Remember they were paying MUFON for case report information from the CMS database. We agreed from the beginning that they would not share their research findings or test results with us.

On a personal note, if I had known about the recovered crash wreckage back then, I would have been on my knees begging to cross over to their side of the mirror.

Seriously though, there was nothing sinister about any part of this deal with MUFON. We worked with these people on a day-to-day basis and from my perspective they kept their part of the deal.

That was all very powerful, and if it would have continued, the UFO research community would have benefited from it greatly.

Final thoughts on the MUFON SIP project

I was the point of contact person between the MUFON team and BAASS, meaning that I was the only person at MUFON who had direct access and daily contact with the individuals working at BAASS. As we progressed with the program up and running, we met on a conference call every Friday afternoon. I facilitated the call and sometimes Chuck Modlin from MUFON would be on the call with me. The project manager and several of the team members at BAASS participated in the call. We typically would review the significant reportable events and talk about the cases we investigated that week.

Early on we developed a particularly good working relationship. Although we know now, on their side of the mirror they were funded by AATIP through the Pentagon They never really tried to hide that, just saying the funding came from some agency that was not disclosed.

I really respected these guys and trusted them. If I asked a question, they would give me a straight answer, or they would tell you “we cannot talk about that”. But I do not believe they ever lied to me.

Skinwalker Ranch

During the time when the SIP Project was operating, Bigelow owned the Skinwalker ranch. There has been a lot of public interest in the extremely high strangeness associated with this mysterious place.

A team known as NIDS (National Institute for Discovery Sciences) was commissioned and financed in 1995 by Robert Bigelow. He set NIDS up to research and study various fringe science and paranormal topics, most notably ufology. Some very distinguished scientists and PhDs were a part of the original NIDS team, which was headed up by Colm Kelleher.

When Bigelow purchased the ranch, NIDS was used to study the phenomenon and the very strange events happening at the ranch.

Many of these same people later worked behind the scenes at BAASS, including Colm who headed up the team at

BAASS. I knew him and communicated with him regularly during the SIP project. He was a brilliant and distinguished individual who was always very courteous, professional and honest with me, I considered him to be a friend.

I understood that some of the BAASS team members were spending time conducting research at the Ranch with various scientific instruments. The feeling I got from some of the investigators was that it was a really creepy place. I know the property was under heavy guard (by armed security people) and very secure. I have never been there but looking back I could probably have obtained permission to visit, but it never happened.

I had not read the book that Colm Kelleher and George Knapp wrote until well after the program ended. If I read that book without knowing those guys, I would have had a hard time believing some of it, but the people involved, I know to be highly credible and I am quite sure it's exactly like they laid it out in the book.

MUFON

The Sinister Side of the Mirror

The Sip Project had been operating for about 10 months when James Carrion resigned as International Director of MUFON, in December 2009, which was extremely unfortunate (that is all I'm going to say about that).

After James resigned, working with MUFON was much more of a challenge. I had answered directly to James and he answered to the MUFON Board. He pretty much kept me insulated from most of the conflict and internal strife at MUFON, so I was free to work with BAASS on the SIP Project. He had my back most of the time. With James out of the picture, it became difficult to deal with the MUFON higher-ups.

The longer things went on, the worse it got for me. I had a particularly good relationship with the BAASS Team and honestly, I know that several (not all) of the MUFON board members were very uncomfortable with that.

Hidden Cases

Toward the end of the program, issues about cases that were mysteriously disappearing overnight started to surface. The Chief Dispatcher on the BAASS/MUFON SIP project, Joe DeEduardo who was also a regional CMS administrator had contacted me on several occasions about high profile extremely interesting cases that were viewed by the Dispatch team overnight, but somehow the following morning the cases disappeared from the system. The team functioned 24 hours but due to time zones there were four hours at night when no one was on duty and the cases were disappearing during that dark time.

Joe had actually printed some of the case reports and emailed them to me. I remember one in particular was an incredible case report about a close encounter and landing on a Native American Reservation in Arizona. There was allegedly videotape taken by reservation police and a number of witnesses to the event including several reservation police officers. The report was verified in the late hours of the night and the next morning it was gone.

I had what I thought was an all-access password to the CMS system and I spent a considerable amount of time looking around in there trying to figure out what happened (where the case went). Ultimately, I found evidence of it in what appeared to be a secret part of the file system.

I was genuinely concerned and irritated about this. I spoke with BAASS, and they did know anything about these cases, but they were also very concerned when they found out that cases had been hidden from us.

I remember having a very heated exchange with several of the higher-ups at MUFON during a conference call. They initially pretended like they did not know anything about it, but I persisted with the copy of the report Joe had sent me and evidence that I had found in CMS. I kept interrogating them until finally, one of them admitted that they removed the case and put it in the locked file that was hidden from BAASS and me. Needless to say, that my relationship with these two guys was deteriorating rapidly.

Financial Issues that lead to the demise of the SIP Project. I did not have responsibility for the financial management and did not have direct access to the bank accounts or accounting. Most of this was facilitated by the Operations Manager.

While doing some of the research for this book, I have run across several articles about BAASS. Interesting that some of the writers imply that somehow BAASS or Bigelow was not satisfied with what they were getting form the MUFON Sip Project.

Most of these issues were concerns raised by BAASS, that they were paying for items (scheduled in the contract agreement) that were not completed by MUFON. Out of the $56,000 paid monthly by BAASS, we were spending about $30,000 on the SIP Project each month.

The remaining $26,000 of the funds being paid to MUFON each month were supposed to be used for the following:

- Special Rewards Program (bonus)
- Incentive Bonus for SIP Investigators ($1500)
- Incentive Bonus for Dispatchers ($750)
- Advertising MUFON on National Media
- Advertising to Police departments across the USA
- Paying for Google ad status to put MUFON on top in searches for "Reporting UFOs."
- Hire a paid Public Relations Officer

None of this was completed by MUFON with the exception that a paid public relations officer was finally hired around the last month when the program ended.

The Dirty Laundry

MUFON was basically sandbagging the $26,000 per month and by the end of June 2009, they had accumulated unspent allocations of about $120,000 from these funds. BAASS wanted the other elements of the contract (noted on the previous page) to be fulfilled and tried to work it out with MUFON.

In all fairness MUFON was a relatively small Non-profit organization at that time and had never received funding like that coming in before. In my assessment the MUFON Board was probably a little overwhelmed with everything that was coming at them, which explains in some part why they were hesitant in fulfilling some elements of the contract as mentioned.

I remember one of the MUFON Board members was extremely belligerent with the BAASS team on the resulting conference calls. He maintained MUFON's right to keep the funds and did not want BAASS telling them at MUFON what to do. I was on some of those calls, and it was embarrassing to say the least (The expression biting the hand that feeds you, comes to mind).

In the period from July 1st through October 31st, BAASS cut back the monthly payment from $56,000 to $25,500, because in their view the other elements of the contract were not being fulfilled, due to disagreement about how the remaining funds were to be used. They were still hoping to resolve the financial issues, which did not happen, and subsequently all the funding being paid to MUFON was then stopped by BAASS. As indicated, we were spending over 30,000 each month on the Sip Project which continued to operate through the first week in January 2010, by consuming the remaining funds MUFON had left.

At the end of December 2009 two members of the MUFON Board ordered me to delete the passwords for the BAASS team members and block their access to CMS System. I contacted the BAASS team and advised them prior to blocking their access, which was a very sad conference call to say the least.

I was instructed to schedule a conference call meeting with the Dispatch team members and advise them that their jobs were being terminated immediately. The employment of the Operations Manager, Public Relations Officer and Myself was terminated a few days later, which resulted in the close of the SIP project.

After all the dust settled by February MUFON had only about 15,000 remaining in the bank. They eventually closed the office in Fort Collins and laid off the last two remaining office employees.

MUFON ultimately left over half of the allocated SIP Project funding on the table as the Contract Agreement originally provided to pay the amount of $672,000 in the first year to MUFON.

No Axe to Grind with anyone at MUFON

When the Sip project ended so abruptly, I was extremely disappointed along with many of us who worked on the project. The guys on the BAASS team were disappointed as well because we all knew that this could have gone on for another 4 or 5 more years.

Currently, I have no animosity toward anyone at MUFON. As a point of clarification when referring to the MUFON Board in this book, there were several Board members I felt were responsible for the financial issues mentioned above, that led to the end of the program. These individuals are no longer on the MUFON Board and have left the MUFON organization.

The most difficult person I had to deal with, who I felt was directly responsible for causing the premature end to the BAASS MUFON Sip project, recently was forced to resigned as International Director of MUFON.

I can say with certainty that the BAASS team was very satisfied with the investigative work we were doing and the information that we were submitting to them each month.

Project Ended

After the BAASS/MUFON SIP Project ended, I published two articles on my blog **"from the Dark Side of the Looking Glass"** about what caused the failure of the BAASS / MUFON SIP Program. My blog received about 14,000 views. I received many supportive emails about my writing from a variety of MUFON members and other distinguished individuals involved in Ufology, some of which actually surprised me.

Some excerpts from those articles were included in this chapter and it is fair to say that the information published in 2011 on the blog has been publicly viewed, read, and disseminated widely for a number of years. You will find a link to my blog and can read the articles, in the references chapter at the end of the book.

In May 2017, 60 Minutes interviewed Robert Bigelow.
CBS 60 Minutes broadcast an interview with Robert Bigelow in May of 2017 which was titled *"Bigelow Aerospace founder says commercial world will lead in space".* [15]

During the CBS 60 Minutes interview when Mr. Bigelow was asked by correspondent Lara Logan, "Do you also believe that UFOs have come to Earth?" he responded, "There has been and is an existing presence, an ET presence" . . . and "I spent millions and millions, I probably spent more as an individual than anybody else in the United States has ever spent on this subject."

He also indicated the FAA referred reports of UFOs and other unexplained phenomena to a company Bigelow owns.

Note these FAA UFO reports were sent to BAASS and at that time I was aware of the process that BAASS used with the FAA. At one point I wanted to apply to the FAA to get the records also sent to MUFON, but it never happened. We were not privy to any of the information the FAA sent to BAASS.

KEITH BASTERFIELD - Researcher [16]
My friend Keith Basterfield who publishes an excellent UFO Research Blog in Australia, contacted me in 2019 via email and posed a series of questions about BAASS as follows: Questions about the names of the individuals who were my

main liaisons at BAASS and how frequently, I interacted with these individuals?

- Also questions about what they shared with me regarding their personal thoughts on the nature of the UFO phenomenon?
- Question about what stage did I learn that BAASS was partially funded by the DIA?
- Questions as to details of any cases which BAASS investigated outside of the 69 reports listed?

Lang responded by email, as follows:

1. "I learned about the DIA funding when the articles broke in the New York Times and Washington Post this past year.
2. As indicated in my blog posts, the BAASS MUFON SIP project was essentially a contract by BAASS where BAASS purchased information from MUFON (sighting reports and investigation reports.)
3. As you know, I supervised the STAR Team dispatchers and investigators during this SIP project. The dispatch team reviewed the daily MUFON UFO sighting reports looking for the more interesting and credible cases, which were provided to me. (Out of approximately 800 monthly UFO sighting reports, which were submitted to MUFON, approximately 35 to 45 of them would be selected.)
4. I reviewed the cases and assigned investigators to travel to the locations, evaluate information, interview witnesses and write reports.
5. I edited the final case investigation reports and forwarded them to BAASS.

6. BAASS was a proprietary organization, and they did not share any of their research with me during that time. I had complete discretion as to what to investigate, which investigators were to be deployed, where and when resources were allocated for cases.
7. An operation person arranged for airline flights, travel reservations, hotel accommodation and pay for investigators. **I was the only person at MUFON who had direct access, and daily contact with the individuals working at BAASS**. I say so only because after the project ended there were a number of individuals (that had no part in the project) who came out on radio shows claiming to have sinister inside information about the project, which was completely false.
8. For the record (on the record) I can tell you that those individuals I worked with at BAASS were honorable, highly credible and always treated me with respect. We had a particularly good working relationship.
9. I sincerely believe that BAASS fulfilled the obligations and elements of the contract arrangements with MUFON to the letter.
10. Out of respect for those individuals that I worked with at BAASS, I am not comfortable talking about them, and will not share any personal information that I know about any of them."

Blue Blurry Lines - March2020

In March 2020 an article which was written by Roger Glassel and Curt Collins, titled *"The Pentagon UFO Program's*

Secret Partner" was published on a blog site known as: Blue Blurry Lines, that surfaced and found its way to me.

The content is apparently in part the result of leaked documents from MUFON. I am named and quoted numerous times within the article and my picture is on the cover along with most of the other players. Also, most of the PowerPoints that are mentioned were created by me.

I learned a few things from this article, and I can probably authenticate most of it. This article made it a lot easier to connect the dots. I had access to most of the information systems and databases at MUFON and I can assure you that I got my share of breadcrumbs that were left for me to find. I found the Glassel and Collins article extremely interesting and worth reading. Links to both can be found in the references at the end of the book [17]

END CHAPTER

Chapter 5

Disclosure
Government Cover-up

Any discussion about UFOs probably needs to include something about disclosure as a reference point.

Disclosure

Many believe that there is a point in time soon coming when the governments around the world will disclose in some form what is known about UFO/UAP phenomena and acknowledge the existence of extraterrestrials. Clearly the secrecy gap surrounding government activities is beginning to narrow due in part to the internet, communication technology, general increase in knowledge, and an elevation in human consciousness. We will talk more about current news events relating to disclosure that are developing in Chapter 15.

Short Version

The short story is that during the late1930s and 1940s some of the governments around the world led by the United States and Germany recovered crashed UFOs and took the wreckage to secret military locations. The crashes at Roswell and dozens of other locations supplied plenty of UFO wreckage and even extraterrestrial bodies for study. They used back engineering techniques and conducted secret research which originated the development of incredible technology.

Many believe the Nazi scientists surpassed the Americans in the development of advance aerospace technology. Some speculate that a group of Nazi scientists developed a breakaway civilization during early 1950s and they may have actually worked with military industrial and multinational technological corporations.

Operation Paper Clip

At the end of World War II, the American military in occupation of Nazi Germany took over the Nazi research facilities in Germany. They gathered up many of the Nazi scientists and confronted them with an offer to be repatriated to the United Stated and work in what would be NASA as well as other research facilities. This was officially known as Operation Paper Clip. During the late 1940s, 1950s and 1960s the United Stated continued numerous back engineering projects and accelerated the technological research using these individuals.

Truman Administration and UFOs

Harry Truman became President of the United States upon the death of Franklin D. Roosevelt, in 1945. He had been Vice President of the United States for only 82 days when he succeeded to the presidency. During his first year in office, Truman approved the use of atomic bombs on Hiroshima and Nagasaki, which caused the surrender of Japan and led to the end of World War II.

It was the testing of the atomic bombs and the destruction of the Japanese cities that many believe attracted the attention of those from out of town (off planet). More than a few UFO crashes occurred on American soil. (listed in Chapter 15) Research indicates that powerful anti-aircraft radar that was used in the early 1940s during world war II era caused the anti-gravity propulsion systems to malfunction and result in UFO crashes.

James Forrestal and Majestic Twelve

In the late 1940s James Forrestal was Secretary of the Navy and was managing UFO projects for the Truman administration. In 1947, President Truman appointed James Forrestal to be the first US Secretary of Defense, several weeks after the Roswell UFO crashes.

The day after the Roswell crash, New Mexico Senator Dennis Chavez had a private meeting with President Truman. After talking with Truman, he personally called the media outlets in New Mexico and threatened to pull their broadcasting licenses if they continued to report the crashed UFO story.

In the wake of the Roswell recovery operations, during meetings in the Oval Office, an undertaking called Majestic Twelve (MJ-12) became operational to oversee the Top-Secret Research dealing with the UFO issues acting as an intelligence gathering agency, which ultimately controlled the reverse-engineering programs in the 1950s and 1960s. Truman issued a memorandum instructing Forrestal to begin funding MJ12 shortly after these Oval Office meetings.

Majestic Twelve (MJ-12) included members of the Council on Foreign Relations. There were 12 members of this group made up of the top officers of government, military and directors of the Council on Foreign Relations.

In early 1948 another UFO crash occurred in Aztec, New Mexico. Forrestal was involved in the UFO crash retrieval. These retrieval operations involved a number of alien bodies and wreckage of extraterrestrial vehicles.

Privately his aides started noticing that Forrestal was developing some nervous mannerisms. This could have been because he was becoming more uncomfortable with the secrecy in hiding all this from Congress (and the public). There is no doubt that he could have been deeply disturbed by some of the things he was exposed to during crash retrieval.

He conveyed his concerns about the covert operations to Truman and wanted to brief some of Truman's political adversaries about the UFO situation, which caused Truman not to trust him. There is intelligence indicating that at one point Forrestal may have threatened to talk to the media and press.

In any event, President Truman started to ease him out of office in early 1949, because of his alleged declining mental health.

He was reportedly shattered when Truman abruptly asked for his resignation. He was diagnosed with "severe depression" and ended up at Bethesda Naval Hospital in

Maryland, where he fell to his death after going out of a window in his hospital room on the 16th floor.

Eisenhower administration

1953 was Eisenhower's his first year in office. There were 10 more alien crashed disks that were recovered along with 26 dead and 4 live aliens.

Early in 1953 Eisenhower turned a member of the Council on Foreign relations, Nelson Rockefeller for help with the alien dilemma. Rockefeller began coordinating the secret structure for control of the UFO situation by using MJ-12 and ultimately taking control out of the hands of Congress and the President.

During the Eisenhower administration, the US government began to utilize technology corporations and aerospace firms to further develop the research.

In a book written by US Army Colonel Philip Corso, he explains how he was assigned to a secret government program in the Pentagon, that provided material recovered from extraterrestrial spacecraft crash wreckage to private industry to be used in reverse engineering programs. [18]

Colonel Corso outlined in the book many of the specific companies that the Pentagon shared the crash wreckage material with, and how they developed technology from it.

For example, he named Bell Laboratories (now AT&T) who worked on fiber optic technology from tubing that was first recovered from wreckage. DuPont Laboratories developed Kevlar from material given to them by the Pentagon. IBM began developing the first micro-chip technology from recovered microcircuits, just to name a few.

Remember back in the late 1950s electronic technology was carbon based (think of you grandparent's old radios with the big tubes inside) and then overnight by the 1960s the transistor age evolved, which is Silicon Technology.

These clandestine alliances were made through MJ-12 and money was funneled into classified back engineering projects. Deals were made to allow the companies to keep patent rights and make unbelievable profits as long as the technology was made available for government and military applications. If you don't believe anything in Colonel Corso's book, just look at the companies he talked about from back in the late 1950s and you will see these corporations are now the multi-billion-dollar blue chip corporations of today.

President Eisenhower initially trusted Rockefeller. However, looking back, empowering Rockefeller was probably the biggest mistake Eisenhower made for the future of the United States and almost certainly all of humanity.

"Beware of the Military Industrial Complex!"

By the end of President Eisenhower's administration, the Federal government's control in this research was diminishing thanks to Rockefeller and MJ12. Control was predominately in the hands of private multi-national contract operatives, global corporate technology, and aerospace firms, which President Eisenhower referred to as Military Industrial Complex. When he left office, he warned us about their power and control in that famous speech "Beware of the Military Industrial Complex".

Eisenhower threatened to invade Area 51

There is an incredible video that was produced by Richard Dolan who interviewed an elderly gentleman who worked for CIA and was part of Project Blue Book. He had a "Q Clearance", which is the highest Security Clearance in the United Stated that is well above TOP SECRET. He claimed to have been with President Eisenhower in several meetings about Area 51, the Groom Lake testing facility, and at S4.

As part of this elderly gentleman's work in Project Blue Book, he described how they met with witnesses and tried to discredit them, smiling and saying something about Swamp Gas theories. In the video he describes how he was dispatched to the oval office with his boss to meet with Eisenhower and Nixon.

Eisenhower was concerned that the operations at Area 51 and S4 were being kept from him and Congress. The elderly gentleman describes how Eisenhower instructed him and his boss to go to Area 51 and deliver a message. To tell them they had a week to brief the president on the operations at the facility or Eisenhower threatened come to Colorado, take control of the first army and invade the base.

The gentleman describes in detail how he and his boss went to Area 51 and then were escorted to S4 where they were shown alien bodies and saucer crafts that used reverse gravitational propulsion. Eisenhower let them use his Lockheed Electra (which was essentially the first Airforce one). They returned to the White House and reported to Eisenhower who was very worried about how the Black Projects had grown so out of control.

The man is elderly and in bad health but very believable and worth watching. A link to Richard Dolan's video can be found in the references at the end of the book [19]

JFK and UFOs

President Kennedy was a Naval Intelligence officer during his military career. Research indicates that his father Joseph Kennedy Sr. who was a very wealthy powerful businessman and American politician, contacted his friend James Forrestal and asked him to take Jack (JFK) under his wing and help to groom him for a possible political career. A long-term alliance developed between JFK and Forestall. JFK was in Germany in the aftermath of the war and research indicates that by then he had a close relationship with James Forestall who was Secretary of the Navy at that time.

Form my experience in the early days, Naval Intelligence was the main player in the UFO government research. It is very likely that Kennedy knew firsthand about the Germans Nazi UFO research.

Kennedy served in the House of Representatives from 1947 to 1953. Then later as a US Senator from 1953 until his election in 1960 to President of the United States.

Because of his Naval Intelligence background and the fact that he had been privy to the UFO research by the Nazis, he was one of the US Congressmen trusted with the Roswell secrets. Some sources indicate that thanks to James Forrestal he was actually invited to be there during the recovery operation.

When Kennedy succeeded Eisenhower as president he was equally concerned about the Military Industrial Complex and Black Projects that had come to be out of control.

Historical accounts indicate that JFK and Eisenhower met secretly in the White House on a number of occasions for dinner and cocktails, where they were able to candidly discuss their concerns.

Kennedy, being prompted by Eisenhower, was obviously very concerned about the operations at Area 51 and S4. History tells us that his inquiries were stonewalled in the same way that Eisenhower had been treated.

The President of United States is the Commander in Chief of all military services and bases, he can visit any military base at any time desired. It is interesting to note that Area 51 and S4 are not U.S. military bases, but rather secret facilities controlled primarily by the CIA. It is pretty well documented the President Kennedy had been to the MacDill Airforce base in Florida, where alien bodies and crafts were apparently stored.

After the Cuban Missile Crisis, JFK had great concern that a UFO event might be misinterpreted as a missile strike and cause a nuclear war. He had apparently been communicating with Khrushchev in a consorted effort to share information about UFOs with the Soviet Union, to prevent a potential nuclear disaster that might be triggered by a UFO.

His contempt for the CIA, in his promise to break the CIA into a thousand pieces and the communication with USSR about UFOS probably caused them to terminate his life with regrets. The fiasco at Dealey plaza in Dallas was most likely orchestrated by the CIA.

Interestingly the bodies of both JFK and his mentor James Forrestal ended up being autopsied at the same place, Bethesda Naval Hospital in Maryland.

Secrecy - Covert Means

It has been said that the secrecy surrounding the UFO/UAP phenomena and the covert back engineering programs have been classified higher than the research for the development of the Atomic Bomb.

Project Blue Book

Project Blue Book was one of a series of systematic studies of UFOs conducted by the United States Air Force. It started in 1952, and it was the third study of its kind. The first two were projects Sign (1947) and Grudge (1949).

We have the benefit of 50 years of history and the stories of investigators associated with Project Blue Book, who interviewed UFO witnesses, confiscated their photos and did the best they could to discredit and intimidate them. It seems that government or contract operatives in keeping with the need for secrecy were conducting covert surveillance on witnesses, investigators, and experiencers, in order to gauge the extent to which the public was aware of the UFO/UAP phenomena and the existence of extraterrestrials. Project Blue Book ended in December 1969. [20]

UFO Organizations, Reports and Data Bases

It is interesting to note that some of the grass roots UFO organizations started to solicit reports for UFO sightings in 1970. **Is it a coincidence that organizations like MUFON started at almost the exact same time that Project Blue Book ended?** I have often wondered if this was a new way for government agencies to use UFO Organizations to monitor the public awareness about UFO activity, without the extreme expense associated with covert surveillance of witnesses and investigators.

Not to mention the plausible deniability aspect which allows government officials in a formal chain of command to deny knowledge or responsibility for any actions committed and that private organizations are not subject to FOIA (Freedom of Information Act) requests.

NORAD

The North American Aerospace Defense Command safeguards the airspaces of the United States and Canada by monitoring unknown and unauthorized air activity approaching and operating within our airspaces. [21]

NORAD monitors inner space so carefully that a small bolt or piece of debris from a satellite is scrutinized and cataloged. There is nothing that enters the atmosphere of the earth that is not detected! There is nothing that materializes in the airspace above the earth that is not detected! [21]

Therefore, any UFO/UAP event (sighting) that occurs is likely known and monitored by NORAD. Some are investigated and sometimes intercepted by military aircraft.

This explains why so many sighting reports indicate that witnesses observe a UFO that is followed by Fighter Jets or Black helicopters.

If that is so, then the question arises as to why the government would be interested in UFO sightings, witnesses and civilian investigations, when if in fact these agencies know exactly what is happening?

The most important element of a UFO organization is the database. In the early days, as indicated UFO Investigation reports were typed on paper and case reports were submitted in manila envelopes that were sent through the US mail. These reports were not generally available to the public.
With the development of internet databases and automated reporting sites on the internet, these computerized case management systems (CMS) like the one used by MUFON evolved. Individuals that have a sighting or experience can go online and submit a report.

These report websites have evolved to be quite complex and allow the individual to report in detail the experience, by answering questions and filling in blanks on the report form. Digital images, audio recordings and video can be uploaded as well. These online reports feed into a database that can be sorted for a variety of reporting applications. As the database grows over the years the data becomes more useful and valuable.

MUFON Database, who has the keys to the back door?

In the case of MUFON CMS, limited access to the data is also made available to members of the organization. The design and maintenance for the servers of such an extensive computer system would certainly be very costly. During my time with MUFON, I was never able to understand how this massive system was financed. There were no records of payments for the database, and no one could explain how it was paid for.

Some of us who have had unlimited access to these reporting system databases, speculate that they are likely designed by very sophisticated technology corporations like IBM and government contractors. On the surface, these databases appear very legitimate to the user, as an individual submitting a report, or the investigator using the system. But hidden from public view are complex back doors and control systems which allow those who design these systems unlimited access to utilize and sort data, as well as conceal extremely sensitive UFO reports. Irrespective of what is being done behind the scenes many of the field investigators and grassroots members still benefit greatly from the experiences that they have investigating cases from the CMS system.

GPS Coordinates for Investigative reports

Another thing which seems strange is some UFO organizations, specifically MUFON are now requiring GPS Coordinates for all Investigative reports. This was something that we started during the BAASS MUFON SIP Project. We were required to include GPS Coordinates (latitude and longitude) with exact times and dates of the event, on all reports submitted to BAASS.

Regressive Satellite Imaging

No one at BAASS ever said so, but I kind of figured that a multi-billion-dollar Aerospace company like Bigelow might have access to some satellite technology, where they could use regressive imaging (knowing the date, time and coordinates of an event, you can go back and review recorded images from satellites that orbit the earth and photograph the surface). To my knowledge no UFO organizations have access to regressive imaging satellite technology, but sophisticated technology corporations and government agencies do, which makes an interesting case for questioning the requirement for GPS coordinates.

Going back to the idea of monitoring the public at large. To know what the public is seeing, perceiving, photographing as well as who is communicating on any level with whatever extraterrestrial or inter-dimensional entity, is unbelievably valuable information. Also, what is NOT seen or perceived by the public may be even more valuable.

Very High Technology

Fast forward 50 years and realize that these technology corporations and aerospace firms operate outside the control of any government, including the United States (that originally funded them and provided alien technology to them). In secret, they have developed advanced technology that is 100 years ahead of anything we currently have in the public domain and have made trillions of dollars for their own proposes. They maintain top-secret underground and undersea bases across this planet as well as facilities off planet. Yes, I really said that!

They completely control the news media, the movie industry, and television broadcasting corporations, which will be discussed in more detail chapter 14.

There is a lot more that could be written but not in this book, because I said this is the short story for now.

My point with this short summary is to help the reader understand that Disclosure is not so much about UFOs, Aliens and Extraterrestrials. It is about unwrapping of the secrecy surrounding the Military Industrial Complex and the advanced technology that has been developed, with black project money used at the expense of US taxpayers. Advance Technology associated with free energy production, anti-gravity transportation (vehicles) and medical advances are not available to the public, whose tax dollars paid for the research.

The Big Picture

There are clearly alien and extraterrestrial vehicles that visit this planet, piloted by different species from different places in the galaxy, from different dimensions and different times. Also, terrestrial technology that is tested and utilized by the aerospace and the military industrial contractors. Many of the triangle shaped objects are thought to be in this category.

Going even deeper down the rabbit hole, there is evidence that efforts with Aliens in corporation with Aerospace Contractors are being conducted in what is known as the Secret Space Program. Before you laugh, GOOGLE "Secret Space Program" and put your seat belt on.

United States Space Force

In 2019, the U.S. Congress authorized the creation of the United States Space Force, which was officially established on Dec. 20, 2019.

The National Defense Authorization Act ultimately facilitated the creation of the United States Space Force which has immense powers and authority not to mention weapons that are yet to be heard of in the public domain.

The commanding General of the United States Space Force sits with the Joint Chiefs of Staff and is among the most powerful Generals.

Essentially the United States Space Force oversees a Secret Space Program with operations off planet that involve humans as well as alien races which has been going on for a long time.

The space force was created in part to ultimately legitimize the Secret Space Program and bring it into the public domain as part of the long-term disclosure project.

In near future we're going to hear a lot more about the United States Space Force.

I have often wondered if this was the beginning in a series of steps which will eventually lead to public disclosure about the current activities in space conducted by the United States. More about Disclosure and what might be coming soon in Chapter 15.

END CHAPTER

Chapter 6

Government Contractors - Police - Military

Prior to 911 American law enforcement agencies often worked independent of each other. They used to refer to this concept as a silo, because very often each government agency was kind of doing their own thing. The FBI and the local city police department might have been investigating the same bad guys and not ever consulted with each other.

After 911 and the destruction of the world trade centers in New York, the U.S. Department of Homeland Security (DHS) was organized. Part of the underlying purpose in organizing DHS was to connect all the law enforcement silos together. There is great power in getting all the government agencies to work together and communicate with each other, which was the task at hand.

I am proud to say I was a part of all that being one of the original employees who stood up (started) DHS. As indicated, I was assigned to TSA and my job was essentially that of a federal liaison. I worked with airport authorities, airlines and all the law enforcement agencies that had responsibilities at the Airport.

We had meetings in the local Federal buildings where the various law enforcement agency heads would meet. FBI, ATF, ICE, Sheriff, Chief of Police, Airport Police and even the managers of the Air Traffic Control Centers. I was often a key speaker at these groups and represented TSA.

Dealing with Police

Police officers are like anyone else in that some of them are interested in UFO cases. I have found that generally police will help you as an investigator, but the adverse publicity that is associated with these types of cases can really hurt them, career wise.

For example, a deputy that had an experience involving high strangeness when responding to a complaint by a witness, might be very objective and direct in telling you about a high strangeness case. The problem is that his boss, the Sheriff of the County has to run for election every four years. The sheriff is not going to be very comfortable with one of his deputies making some kind of statement about a very strange event that occurred last night, when it gets in the newspaper and ultimately looks bad for the Sheriff.

So, if you are working on a case and the police are involved, you'll probably get a lot further with the police officers or deputies, if you assure them that anything they tell you will be kept in strict confidence, and you need to keep it that way! Because if one of these guys was honest with you and helped you out, the last thing you want to do is cause him problems on the job.

Once upon a time in Virginia a long time ago, I was working on a case where we conducted a night watch. A night watch is where several investigators get together and set up photographic equipment, and video camcorders to watch for something to happen. This case was about an orb or light that apparently frequented a certain area over that mountain. It was observed by a number of witnesses.

I remember we set up video camcorders on tripods watching for the lights to appear when a police car pulled up. The officers walked up to the top of the hill where we were set up. Apparently, a neighbor had called about suspicious activity.

They obviously wanted to know what we were doing. We directly told them that we were investigating anomalous phenomena and were trying to videotape a strange light that seemed to be appearing at the top of this mountain.

We had ID and investigator credentials which really helps in situations like that. Once they realized that we were not burglars scoping out someone's farmhouse, they allowed us to stay and continue our night watch. They actually hung out with us for an hour as so chatting about the phenomena and seemed very interested in what we were doing.

So, in circumstances like that it is best to be honest and direct with police, because from time to time they are going to confront you when you are in the field. Someone is always going to see you and make a suspicious complaint. It is great if you have some kind of investigator ID or Press Credentials with you.

I currently operate a publishing business and we issue press credentials to our reporters and photographers.

Government License Plates

We had a strange situation that occurred on one of the star team investigations. I had deployed in a very experienced female investigator to work on a case in Tennessee. As I remember the location of her interview was about 1½ hour drive from her house. She met with the witness who had taken her to a field and shown her where the sighting and close encounter had occurred.

She noticed that there was a vehicle parked a couple of hundred yards away from the location where the event occurred, and the vehicle was occupied. They noticed that there was a government license plate on the vehicle. After she completed her interview, she went to a local diner to get something to eat and noticed that there were several vehicles parked in the parking lot, including one that had government plates like the one pictured here.

After lunch, she was in her personal vehicle heading back home. She noticed that a vehicle a Ford Crown Victoria, (very popular choice for law enforcement agencies back then) was directly behind her and followed her for the entire 1½ hour drive back to her home. As she was about to take the exit for her home, the vehicle sped up and passed her, pulling directly in front of her, so she could notice it had a government license plate on the car. Then it sped away as she got off the exit ramp to her home.

Generally speaking, you will find government license plates (what we called G-Plates) on vehicles that are owned and operated by the United States government. Federally owned vehicles have G-Plates. The states and commonwealths have their own form of the state government plate. Vehicles with G-Plates can utilize special access like HOV Lanes and can park in spaces that are for authorized vehicles only. Local police and state police officers will usually not pull over a car that has G-Plates on it for minor traffic offenses.

We were concerned about what happened to this woman and I discussed it with the BAASS team. The consensus of opinion was that the vehicles were used to intimidate this person. Interestingly it only happened one time on the SIP Project.

I personally have participated in mobile surveillance operations. I can tell you that if you were the subject and we had you under surveillance, you would never have any clue that you were being watched. Depending on the case we typically would have at least two investigators and sometimes three, all driving very plain Jane looking vehicles. Not to mention that we would be in contact with each other via walkie talkies, so if I followed you on the road for a while and you were about to get off an exit, another investigator would pick you up in a different vehicle and follow you for a while.

We used video cameras that had 450X power zoom capability. I could literally read a newspaper from about 100 yards away. Oh yes, they had night vision capabilities also. If I were sitting in a minivan, parked 100 yards away from you in the neighborhood, with tinted glass, there is no way you could see me with the naked eye at night or in daylight.

Bottom line if you are ever being under surveillance by some government operatives, I can assure you that you will never know it.

Recovery Retrieval Teams

The average UFO/UAP Investigator scrambling to the scene of one of these anomalous events to investigate a landing or crash would typically be in way over his or her head.

It is not likely that a civilian UFO investigator would get to such an event before the retrieval team does. But if you ever do happen to be in the wrong place at the right time and witness such an event here is how the situation might evolve in the event of a crash or landing.

Crash Recovery Retrieval Teams (Recon Units)

There are Crash Recovery Retrieval Teams (Recon Units) that are always on stand-by to deploy to the scenes of such events. These Recon Units may be a combination of military and or government contractors. Contractors can operate covertly with less accountability than the Military. These Recon Units are not necessarily always involved with UFO / UAP events but would also respond to a satellite crash, missile warhead accident, or any type of space debris coming down.

Let's explore a hypothetical situation!
The first order of business would be to set up a perimeter around the area where the event occurs, by blocking streets, railroad lines and interstate highways leading to or around the area. The surrounding airspace would be immediately restricted to all air traffic (including drones).

A cover story would be sent out to the media to explain the perimeter, like a train wreck, plane crash, quarantine for disease, or the accidental release or spill of toxic substances.

Local police would likely be involved and would assist with perimeter control and access points. Police would not typically be involved in the retrieval process and would probably not know the true nature of the event (or what was actually going on inside the perimeter). However, the cover story would be serious enough to justify the use of deadly force to maintain control of the perimeter.

The vehicles pictured on the next page were traveling in a caravan, which included about 15 vehicles at an undisclosed location in Virginia.

The photos which were taken from public view and show how a Recon / Recovery Unit might appear on the highway while in-route to an anomalous event.

These vehicles have high tech satellite microwave receivers and antennas mounted on top and all the vehicles were painted dark blue and appeared to be in bran new condition. Note that police light configurations (flashing Red and Blue) were mounted on all these vehicles.

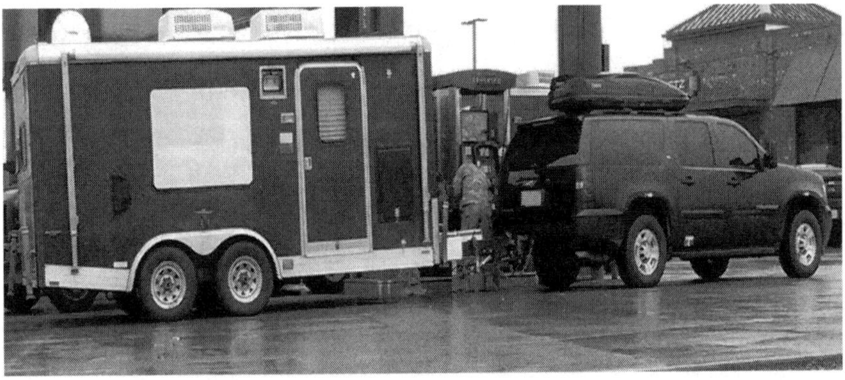

Pictured here is what appears to be a mobile command center and they were all manned and operated by what appeared to be uniformed military officers. All vehicles had Federal Government Plates

In the event of a crash or a landing, Recon Recovery teams would immediately deploy to the scene of an anomalous event.

Warning to UFO Investigators

These guys like the ones pictured on the previous page would immediately take control of the area and are authorized by Federal Law to arrest and detain indefinitely, anyone who confronts them or breaches the security perimeter they set up. Don't even think about sneaking into an area already set up with a security perimeter.

What if you get there first?

Here are some things to consider:

- Do <u>not</u> ever cross a police line!
- Personal safety is most important!

- Time is limited, so do whatever you feel you need to do and leave the area quickly.
- The military, police, or government agents that respond will not be nice to you!
- You can forget about any civil rights you think you have in a place like this. Especially if the military or government security contractors are involved, in which case the rule of law may not apply.
- You could be treated as a terrorist or an enemy combatant with no right to legal representation including the right to remain silent. Not to mention you may not be allowed to contact family members.
- Crash debris could be radioactive, electromagnetic or discharge toxic substances. Note that some satellites have small nuclear reactors which could obviously leak after a crash landing.
- Do not walk under a flying or hovering UAP as the anti-gravity propulsion (electrogravitic) emits a type of radiation, kind of like an electromagnetic field similar to microwave in the way that it will cause tissue damage.
- Do not touch or attempt to touch a UAP that has landed.
- A UAP that is landed on the ground and still powered up can emit ultraviolet radiation from the metallic surface of the craft that will cause burns to the skin like sunburn, also there may be an emission of radiation in the X-ray spectrum that can cause radiation sickness.
- Do not point any kind of high intensity flashlight or high-power spotlight at a hovering UAP! There are known

cases where persons have been burned by bright light rays emanating from UFOs after doing so.

- Psychological effects produced by force fields that could induce a hypnotic state in the observer, including loss of consciousness, memory relapse, and submission to the UFO occupants.
- It goes without saying that any display of firearms or other weapons on your part could be construed as hostile and will most likely end disastrously.
- When you vacate the event site area, you may be stopped by police in the surrounding community. They will question you and will definitely search your vehicle.
- If you have a camera, they are going to look at the pictures and will confiscate the data card (flash drive/ SD card) if not the camera or video camcorder.
- The less you say the better chances you have of getting out of there.
- Your Smart phone is your worst enemy so don't call your spouse or significant other to brag about what you might have just witnessed.
- Police have technology that they use to do surveillance on drug traffickers (targets a specific area) that works very efficiently with smart phones. They can track your movement, listen to your phone conversations and in some cases listen to and see video of you, when you think the phone is turned off. Same with your laptop or I-pad if it has a modem connected to the Internet.
- If you think you are being monitored, turning the smartphone off will not work. They can still turn it on and listen to your calls and see your texts and email. Take the

battery out of the smart phone or leave it in the car or at the hotel.

- Always have a backup battery for your digital camera and video camcorder (batteries have been known to discharge quickly in areas like this)

- You should have some type of data security plan. Keep extra data (SDXC) Cards with you in the little plastic case that they sell with them.

- It is a good practice to take half of your pictures on one card and switch the SDXC cards, then store the used one in the little plastic case and stuff it in your sock inside your boot. In the event, you get stopped or someone takes your camera, you still have some of your pictures. (more about data security in chapter 8)

- If you film or photograph something extraordinary you will want to get it on the internet cloud as soon as possible.

- Uploading to several trusted sources simultaneously on some type of shared cloud space would be preferable.

- Uploading this data should be done as far away from the affected area as possible in that communication will certainly be monitored and cellular phone communication may be jammed or shut down.

- Note: lying to a federal agent is a federal offence punishable with imprisonment!

- Note: In most circumstances, police officers need a warrant to search through your phone or other recording device.

- Even if arrested, the police may not search through the digital contents of your phone, and this goes for photos and videos.
- The police may not generally search a person's cellphone for digital information without a warrant during an arrest.

Information contained herein is not intended to be legal advice, so it is best to consult an attorney for more on this, as laws and restrictions vary from state to state.

END CHAPTER

Chapter 7

Use of Video in Investigation

In our younger days, we all remember the fictional tough cop character "Dirty Harry" brandishing his 44 Magnum handgun and saying, "this is the most powerful hand-held weapon in the world". Back then I probably would have agreed.

Combined with the wisdom that comes with age and my years of investigative experience, I now realize that a handheld video camera is by far more powerful. You can change the course of human history with such a video recording device and an internet upload.

In my assessment, the video camera is the most valuable piece of equipment that any investigator can have and use. Learning how and when to use it while investigating and interviewing witnesses will prove to be invaluable.

For those who remember the Discovery Channel shows, which were referenced in the previous chapters, in the Fayetteville case, we videotaped interviews with many witnesses during the original investigation. It was the video that caught the interest of the producers who ultimately produced the TV show. About 10 percent of the footage in the TV show was video I actually shot during the investigation.

Video Equipment

With that said a look at camera equipment would be in order. The quality of hand-held video equipment has dramatically improved over the last 15 years. Hand-held video cameras (camcorders) were originally equipped to record on a tape cassette. Some of those cameras had reasonable quality video (medium resolution) but editing and transferring data was very cumbersome. Tape video cameras can record digital video images but in order to download and edit the video, it must be transferred in real time as the tape recording plays through.

Video cameras have evolved to incorporate internal hard drives and use high speed flash drives and SDXC cards that can be removed from the camcorder and plugged into the computer for downloading and editing. Image resolution has also improved.

If you are ever thinking your video might be used on a TV show, then you want to have a newer video camera that has the full-color image HD1080p resolution (1920×1080) and a widescreen aspect ratio of 16:9 which is standard in broadcasting.

Most of the newer better cameras have HD1080p resolution capability and can be set to record widescreen aspect ratio of 16:9. They also have the capacity to take still images which will be stored on the SDXC card as JPEG images.

It is advisable to have plug jacks on the camera for microphone and headphones. A tripod is also particularly important to reduce camera shake. Pictured on the next page is my video camera with a directional microphone mounted on top. This type of microphone dramatically improves the sound quality and picks sound up from directly in front of the camera where it is pointed. These microphones help to reduce background noise from behind the camera.

You will also notice earphone headset which is plugged into the camera. This allows the photographer to hear exactly what is being recorded. Distracting background noise, like machines running, which may not be noticed with the ears, can be very loud and undesirable when recorded on the video.

Another useful option is a Lavalier Microphone, which can be clipped to the witness's shirt or blouse which will enhance the sound quality and reduce background noise and echo.

Note: to understand the difference between Standard Definition and High Definition, if you pasted a 5″ by 5″ square card on the screen of a 24-inch monitor, this is the difference you would see between the higher and lower resolution. 1080p (1920×1080) is also known as full HD and is characterized by 1080 horizontal lines of vertical resolution. [22]

Digital Single Lens Reflex (DSLR)
Single Lens Reflex (SLR) 35mm film cameras were originally designed with interchangeable lenses and a system of prisms and mirrors that allowed the photographer to look directly through the lens.

When the photographer pushed the shutter button, the mirror flipped up so the image could be projected onto the film which was exposed by a shutter, that opened allowing the light to expose the film. When the film was exposed, the shutter closed and the mirror dropped back into position to allow viewing through the lens again.

In the evolution to the digital age this same concept has been implemented in redesigning digital cameras to obviously eliminate the need for film using an electronic Image Sensor.

A DSLR camera looks like its 35mm predecessor, except a digital Image Sensor has been put into place where the film was. When the shutter button is pushed, the mirror flips up so the picture can be projected on to an Image Sensor. You see what the lens sees, without any delay and in real time. This is ideal for fast action and low light shooting. An LCD viewing screen is placed on the back of the camera (usually about 2 inches high and 3 inches long). Pictures and video can be viewed on the screen. DSLR cameras are designed with the same lens mounts so that older 35mm lenses (of the same manufacturer) can be used interchangeably without the need for any adapters.

Some of the newer DSLR cameras can shoot cinema-quality full HD video. In the movie mode, the mirror flips up and allows the images to project directly on the image sensor while the monitor screen on the back of the camera displays what is being seen by the lens and recorded.

Be aware that most of the DSLR cameras have Image Sensors that can overheat and may shutdown during a prolonged video recording after only 25 or 30 minutes. When recording longer interviews, try to schedule a few breaks, or use a conventional video camera.

Always have backup batteries and a car charger for your video camera and DSLR batteries. Video camera batteries come in different capacities but most high-capacity camera batteries now available can provide power for 230 minutes, which is over 3.5 hours of continuous operation.

Same for DLSRs, a fully charged battery will be good for about 950 photo shots or approximately 80 minutes of HD video footage. I recommend having two of these batteries for your video camera and two batteries for each DSLR camera body,

DLSR Travel Solution

DLSR cameras may be a travel solution to packing one camera for both video and photography, in a much smaller package.

Pictured here is my DSLR mounted on a compact travel tripod, with an 18 x 55mm zoom lens. Notice the wireless microphone receiver in the hot shoe and digital earbuds for monitoring the sound quality.

One of the advantages of using DLSR cameras, is the wide variety of interchangeable lens and accessories that are available.

I use Nikon DSLR equipment and carry 2 camera bodies. The lenses recommend are as follows:

- 18mm x 55mm zoom lens for general photography.
- 70mm x 300 telephoto zoom for surveillance.
- 10mm x 20mm for wide angle photography.
- 35 mm (normal) for forensic photography.

Most of the newer DSLR cameras utilize an Image Sensor which is about 1/3 smaller than the original 35mm film cameras.

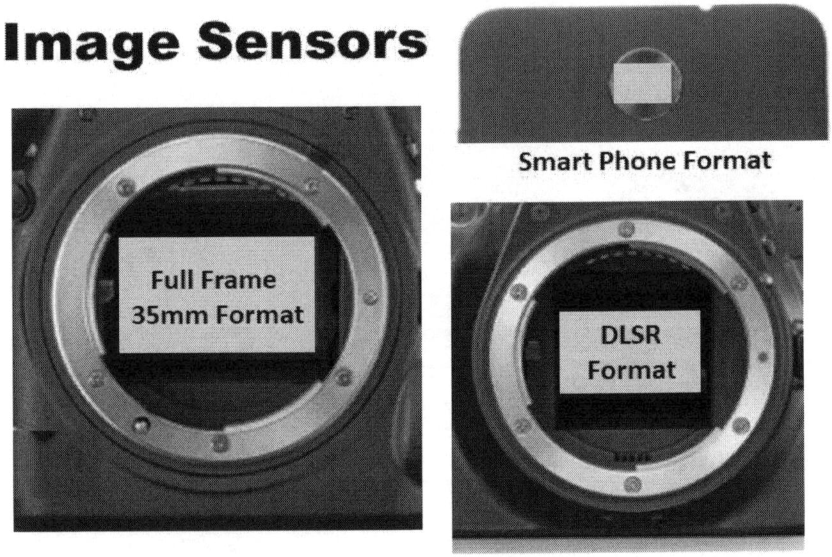

The important thing here is that although the DSLR Image Sensors are smaller they are much more powerful and can generate images in the 20-megapixel range. The good news is the smaller size effectively increases the focal length of the lenses. For example, on a full frame 35mm film camera, a 50mm lens would show what would be seen by the human eye. Photographers have typically called this a “Normal Lens.”

Because the DLSR Format is 1/3 smaller, a Normal Lens would have a focal length of 35mm, which would show what would be seen by a person from the vantage point of where the photograph was taken.

Photos for law enforcement purposes, that are used as forensic evidence in court proceedings are usually required to be taken with Normal lenses.

The same concept Increases zoom power with telephoto lenses, images will appear 1½ times closer with DSLR Cameras. For example, a 70mm x 300 telephoto zoom lens used for surveillance, would have the capacity of 105mm x 450mm effectively increasing the range of the lens.

Nikon produces a speedlight flash unit that incorporates LCD lights on the front of the unit that can stay on for almost an hour, which works great for videotaping at night at close range.

Note that high quality video can be recorded with this type of equipment, however the audio portion of the video is equally important and can be enhanced with the use of a lavalier microphone or a set of wireless microphones. I keep both with my DSLR cameras.

SDXC Cards

Keep extra data SDXC Cards with you in the little plastic case that they sell with them. A 64 GB SDXC card will be good for about 6 hours of recording in MPEG-4 format at 29.97 frame rate in HD1080.

Sony produces a SDXC Card known as TOUGH. These SD cards have very fast recording speeds at 300 MBS and are very tough. They are one-piece construction with no write protection switch and are 18 times stronger than conventional SD cards, not to mention they are waterproof, dustproof, and impervious to higher temperatures.

When recording for extended periods most newer video cameras will break up the file into 30-minute pieces, so for example, if you record a 2-hour meeting, there will be four MPEG-4 files on the SDXC card when you are done. If you paste them into the same file on the computer and play the first one, they will usually link together, and the video will play through uninterrupted.

Video Recording an Interview

The question arises as to when do you try to videotape an interview? **The answer is every chance you get!**

Investigators are hesitant to use video because they do not have confidence that the witness will agree to be videotaped. This is a big mistake because most witnesses will agree to be videotaped if approached properly.

First assure the witness of confidentiality and explain that the video will be used only with their express written permission. It may help to explain that it speeds the interview process in place of writing notes for hours. Use of a written release and authorization form that will document the exact agreement for the use of the video, will make the witness feel more comfortable and you should offer to give them a copy.

There are several especially important advantages to videotaping an interview. First it will document the testimony forever. If the witness dies, changes their mind, or is not available any longer you still have a record, which is much more realistic than a written report.

Video can work almost as well as a lie detector in that people tend to be more truthful or careful about, what they say when filmed, and discrepancies can be easily identified and documented. Video can be reviewed later to corroborate and compare witness statements.

It is best to use a specific list of questions. When someone tells the truth, they will be consistent in their testimony because it reflects their memory. If you interview someone twice, over a period of a few weeks, using the same questions you will see in the later interview if the answers are consistent or major discrepancies are observed.

Video is also very useful in consulting with other Investigators. Viewers can see body language, attitude, temperament, and personality. The video can be uploaded to cloud sites to be shared with team members in other parts of the country.

Interview Questioning

- At the beginning of the interview ask the witness to state their name and address and acknowledge that they know they are being videotaped.
- Start the camera at onset of interview so that you will not miss something important that may be said spontaneously.
- Know as much about the case as possible prior to Video Interview.
- Have a list of desired questions prior to interview.
- Help the witness to relax and get comfortable. Take plenty of time. Most witnesses loosen up after 15-20 minutes on camera.
- Do not lead witness.
- Do not offer information to witness about other cases or witnesses.
- Ask open ended questions, the witness needs to do the talking.
- Try to get the witness to explain in detail (tell me more about that) (help me understand that).
- Ask about sounds, smells, taste, moods, and feelings.

- Repeat some questions from list at end of interview.
- Interview witnesses separately, never allow one witness to be in the room while another witness is being interviewed.

Video Evidence Labeling

Fill out some type of video documentation card (like the one pictured here) that includes the witness's name, location city, state, case information, date, time, coordinates, investigators name and camera information and settings.

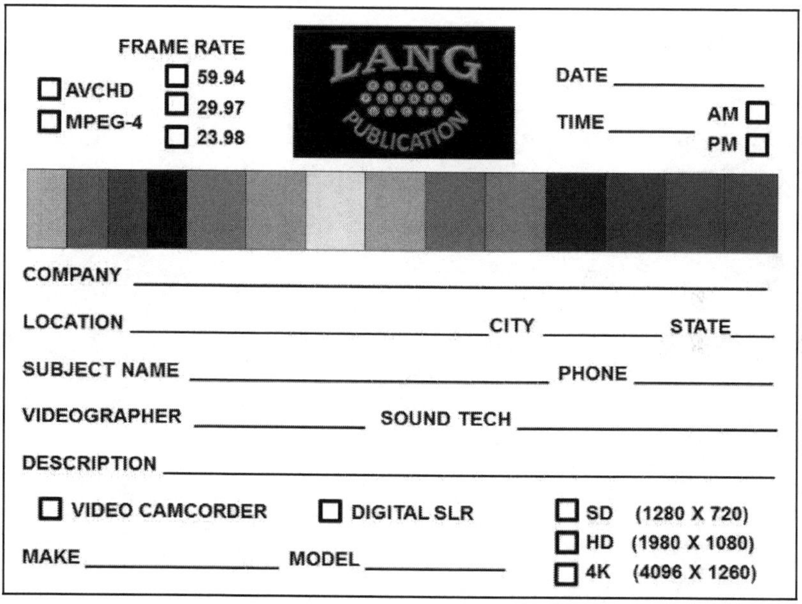
FRAME RATE
☐ AVCHD ☐ MPEG-4
☐ 59.94 ☐ 29.97 ☐ 23.98
LANG PUBLICATION
DATE ____________
TIME ________ AM ☐ PM ☐
COMPANY ____________________
LOCATION ____________________ CITY ________ STATE ___
SUBJECT NAME ____________________ PHONE ____________
VIDEOGRAPHER ____________ SOUND TECH ____________
DESCRIPTION ____________________
☐ VIDEO CAMCORDER ☐ DIGITAL SLR
MAKE ____________ MODEL ____________
☐ SD (1280 X 720)
☐ HD (1980 X 1080)
☐ 4K (4096 X 1260)

At the onset of the interview photograph the card for about 15 seconds. This information will be valuable in organizing and editing your video, after the passage of time or in the event that you conduct numerous interviews. It also enhances credibility with custody and control of video evidence.

You may want to consider photographing the written release and authorization form at the end of the interview for safe keeping on the video. This also documents and directly links the signed release and authorization form with the specific video it was written for.

Camera Placement for Interviews

A room suitable for videotaping an interview should be well lighted and free from distractions. The investigator needs to have control of the environment to limit noise, interruptions, and phone calls. It is advisable to turn off cell phones and HVAC systems that may make a lot of noise. The illustration below shows the layout for camera and seating positions of the witness and interviewer.

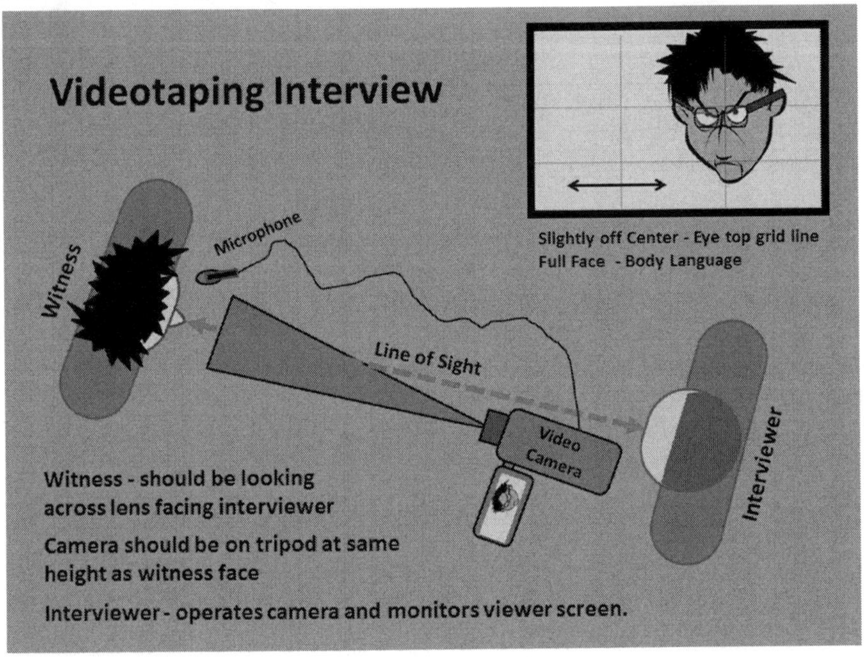

A Lavalier Microphone should be clipped to the witness's shirt and a second microphone could be used for the interviewer. The interviewer may want to use headphones prior to starting the interview, to adjust the volume of the microphones and listen for any unusual noise. As indicated, the HVAC systems may not be noticeable to the ear but may be loud when picked up on the videotape.

One final thought about Setting up the room

Whenever you go to meet with someone always be cautious about the environment you are working in. Especially if someone says "here this is where you sit" or "here this is where you stand". Do not ever let anyone arrange the room, place your equipment or tell you where to sit during an interview when you are videotaping.

Back in the early days, when I first started investigating, we went to a witness's home to conduct an interview. When we came into the house where the interview was to be videotaped, they already had chairs arranged and were very persistent in telling me where to sit during the interview.

During the filming I happened to look up and noticed a small red led light inside a bookcase, which was across the room from where I was sitting. I got up and looked to find a small video camera that was carefully arranged between the books. Fortunately, they were not smart enough to turn off the red led record-light and I was able to discover it. Needless to say, the interview ended badly, but a good lesson was learned.

Videotaping an Event or Object

On more than one occasion the best video and digital camera equipment have been known to malfunction during an anomalous event. There are countless reports of batteries that were fully charged going totally dead during an event when in close proximity to certain types of objects.

Even more interesting are witness accounts where individuals watching a close encounter and trying to videotape an object, report they can clearly see the object, which is in close proximity above them, but when they aim the video camera pointing directly at the object, there is nothing on the video camera screen or digital viewfinder.

One of the best investigators I know of and respect, related a story while speaking at a conference, about an anomalous event he and another investigator experienced. The two investigators were conducting a night watch in an area that was reportedly frequented by an unusual red orb that appeared to be intelligently controlled. They set up a staging area at the end of a large open field and waited.

After several nights of waiting and observation, the red orb vividly appeared and floated over the field in front of the two men as they were watching in amazement. With video camcorder in hand, he indicated that he had pushed the record button and the camera was recording video for over 3 minutes. After the event, they watched the video which showed his feet shuffling around on the ground for 3 minutes. He was so mesmerized by what was occurring before his very eyes that he apparently did not have the presence of mind to point his camcorder up at the object. It happens to even the most experienced investigators.

Multiple Camera Scenario

Sometimes it makes sense to have multiple cameras for such a scenario. Have one video camera mounted on a tripod, set at wide-angle to view the entire area and keep it running continuously. This way if something unexpected occurs spontaneously you have video of the event. A second hand-held camera can be used to focus on any activity that occurs so the operator can zoom in and follow the action. It makes more interesting video if the operator of the hand-held camera is seen in the wide-angle view of the mounted video camera. Having video from two different perspective angles of an object or event, adds credibility and makes it much more difficult to contest or debunk the report.

As far as quick in-field video techniques and shooting on the run goes, the hand-held camera can be utilized in vehicles and outdoor locations. Anytime there are landmarks like road signs or interstate markers it helps to shoot and incorporate them in the video.

A-Roll / B-Roll

"A-Roll" is the primary video and audio of the subject or interview. "B-Roll" is supplemental footage used to visually support or show the background information.

For example, if a witness is interviewed on-sight at the location of an event, which enhances any report, it makes sense after the interview, to pan around the area with the video camera to identify the surrounding area.

When videotaping B-Roll sometimes narration by the investigator while shooting may help to clarify the content by articulating details such as location, date, time and individuals present. The narration can be edited out if necessary but makes documenting and organizing video more efficient.

Remember to always use professional language. Often when individuals are taken by surprise while watching something that is unusual, they use expletive language which can ruin what might be an unbelievably valuable video or at least limit the audience that can view it.

Any time smaller articles are photographed or videotaped, it helps to use some type of scale to allow the viewer to evaluate the size. I keep a set of measuring devices with me in my video gear.

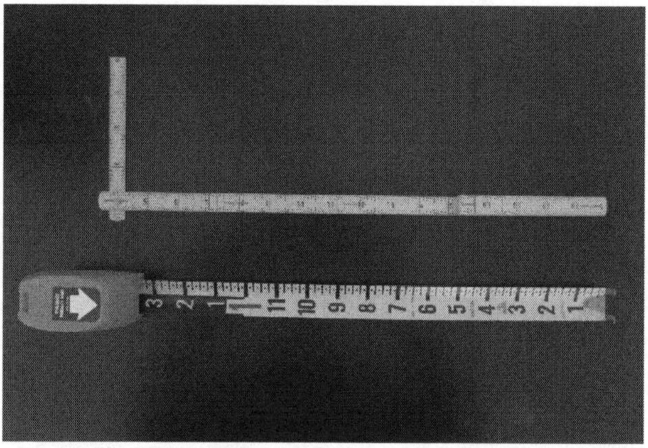

Pictured is a 6-foot large scale tape that opens up rigid like a 6-foot yardstick but rolls up and retracts into the case for storage. The other is a smaller folding ruler.

Video Equipment packing for travel

Traveling to the location of an event or to meet with witnesses involves either automobile or airline travel. Protecting video equipment from damage is best accomplished by using hard sided internally padded cases that are waterproof. We will talk more about travel cases and accessories in the next chapter.

Hotel Editing and Viewing

It goes without saying that if you are videotaping you need to have a laptop with you. Generally, after a day of investigation and or interviewing it is best to go to your hotel (or home if close by) and secure your video. The video will be on SDXC cards in the camera, and it is advisable to download the files onto your laptop by copying the files from the SDXC card to the hard drive in the laptop and a portable hard drive for additional security.

Preview the video files to make sure that the quality and content you expected is in order. It is also advisable to have secure cloud storage available (storage on the internet in a secured server) so that files can be shared and secured in the event of an accident that may destroy the computer or camera.

Editing Video

I use Adobe Photoshop Elements for photographs and Adobe Premiere Elements to edit video. Pictured here is my computer which has been upgraded with extra RAM memory and two screens. It is advisable to have a high-definition monitor like the one pictured here in the middle which is made for editing video. It has a very narrow wide view which lends itself well for video.

Cloud Storage

Some type of cloud storage is essential for storing and sharing video, photos, and documents. Microsoft One Drive or Drop Box are very popular. In the past we used Drop Box and created shared folders.

Basically, a shared file is created and placed on all the other investigators computers. Everyone in the group has the same folder on their computer. Thus, if I have a video file that is 500MB, I simply copy it to the shared file on my computer and the other investigator can view it within a couple of minutes.

END CHAPTER

Chapter 8

Equipment

Most UFO Organizations today are still training investigators in the quest for physical evidence, which is good, Nevertheless I have learned that there is a lot more to all of this as we will explore in the next few chapters.

I am sure that some still think that they will find the answers in the Roswell desert, digging for debris from a crash 75 years ago. Not to say that it did not happen because there is insurmountable evidence that indicates that it did, however any evidence has long since been recovered and hidden from the public view.

I remember one of my fundamental UFO Investigator trainings involved what was called a Radiation Survey on ground, which would be conducted at an area believed to be affected from a landing or close proximity ground effect.

The procedure involved using tent stakes, which were driven into the ground and connected with string in a grid like formation. The idea being that each section would then be tested with a radiation detector for readings, compared with background readings and recorded carefully in the written report.

I carried that bag of tent stakes and string around for 20 years and never had occasion to open the bag and use them. I remember having this same conversation with our friends at BAASS and they pretty much had the same experience, that there was typically not a lot of radiation trace evidence involved in the cases they looked at and investigated.

Radiation Detector

I did use the radiation detector a couple of times but never had any significant readings. Honestly, I kept the radiation detector around primarily for my personal safety. I often thought to myself if it ever starts clicking loud (indication of significant radiation), then I start running away from the affected area, in order to figure out which federal authorities need to be notified.

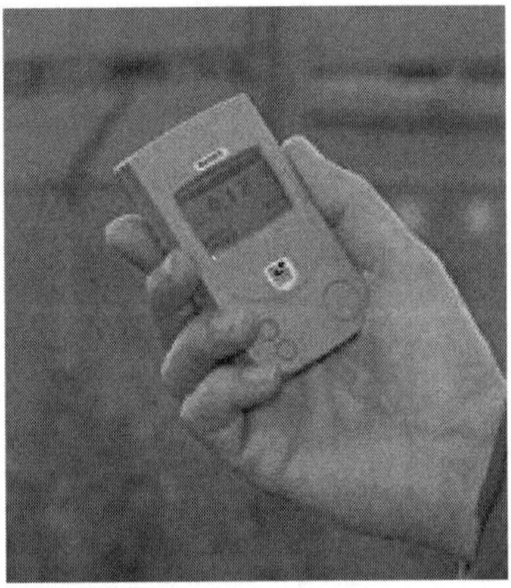

Handheld Dosimeter, Geiger Counter,
High Accuracy Nuclear Radiation Detector

During my time with Homeland Security (DHS) as indicated in Chapter 2, I took classes on Responding to Hazardous Substances, through the Department of Justice (DOJ). They told us that a suggested method to evaluate how far away from an affected hazardous area would be safe, was to extend your hand out with your arm straight and your thumb up. If you cannot cover up the entire scene with your thumb nail, then you are too close. Try it sometime and you will find that is an awfully long distance, so running fast is best.

In today's world, we would define an affected area with GPS coordinates so investigators would be able to pinpoint the exact location of an event at later times.

Note: Radiation surveys should only be done by qualified trained personnel with proper protective clothing and breathing apparatuses.

Vehicle

Your vehicle is probably the most important piece of equipment you have when working on a case during an investigation. If you are deployed on an investigation locally your personal vehicle should be set up similar to what is described in this chapter. If you must travel by aircraft when you get to your destination airport, try to rent a vehicle equipped as described here in.

The vehicle is your mobile studio, workstation and life support system while in the field. A plain looking midsize SUV vehicle in a subtle color that is inconspicuous works best.

The AWD (all-wheel drive) feature is a necessary option, which can facilitate transportation in bad weather (rain or snow) as well as light off-road navigation through fields or on muddy dirt roads. It can get you into where the action is in remote locations, and in some cases get you away from the action if things get dangerous and break bad. It is a place to relax, get cooled off or warmed up while on location. Make sure you have bottled water and a small cooler with some snacks or sandwiches.

There is a substantial compartment in a SUV (behind back seats) that can be used as a workstation and the rear hatch when open can provide shelter from rain, snow and shade from sunlight.

Go-bag Readiness (packed for travel)

Events happen unexpectedly, windows of opportunity open and close quickly and response time is critical, so having gear packed and ready to go is a good idea.

Pictured here is a hard-sided equipment case, a zip tripod case, a wheeled backpack for clothing and a laptop case. Do not forget to include a good water bottle.

Equipment Case with wheels for transportation

A Hard sided padded waterproof case, which is used for basic common-sense equipment that investigators need to carry in the field, will contain "must have gear" ready for "grab and go" to use on location.

You will notice that there is a space or compartment for each piece of equipment and the gear is well organized. If a situation occurs where you must evacuate immediately, by simply looking into the case you will see if there is an empty compartment, which lets you know that a missing flashlight or camera is laying around, which you need to find quickly before packing up and departing.

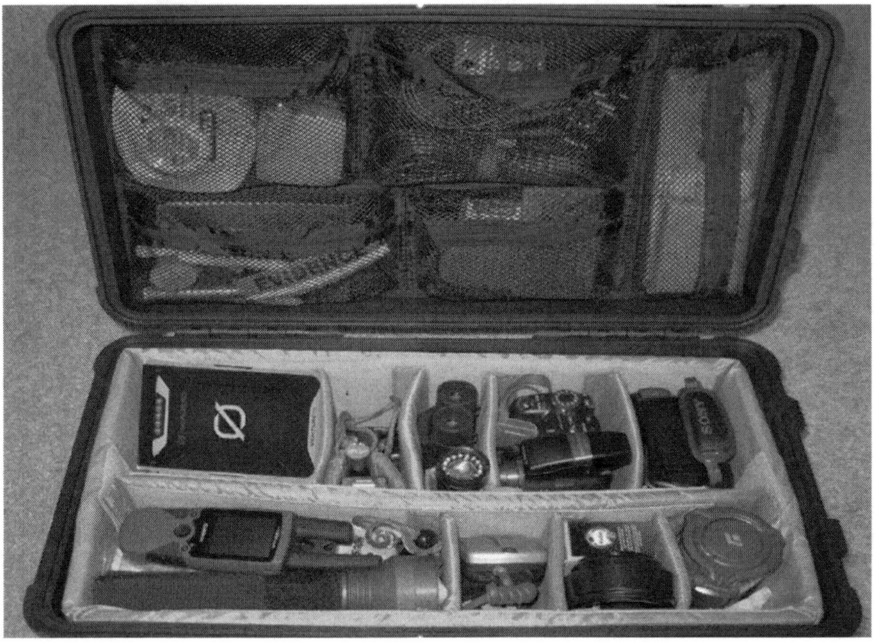

Here is a look at some of the more important equipment that should be carried on a deployment.

- Small video camcorder and digital camera or DLSR which can be used for photographs and video.
- Waterproof Binoculars at least 10 x 42 magnification.
- Flashlight and headlamp with rechargeable batteries
- A battery pack for recharging cell phones and flashlights with built in LED area light.
- Measuring devices 100-foot tape measure, small tape measure and range finder (laser optics)

- Magnification loop 10X
- Star Finder and Color Chart.
- A good walkie talkie which can be used to communicate in areas with no cellular service.
- Handheld GPS units which will be described later in this chapter.
- Area maps and Road maps.
- Topographic Maps.
- Leatherman Multi-tool with scissors.
- Ground cloth.
- 50 Pound Magnet (used with a plastic bag over it to collect ferris metallic properties.

Evidence Collection Kit

Sample containers (plastic bags) disposable polyethylene gloves, evidence collection bags with labels. Sample containers (small plastic jars) ruler, small garden trowel (for fine detail work),

Lights

While serving as a law enforcement officer in an airport environment, of all the equipment I carried, the flashlight was by far the most useful tool. Several types of lights should be packed for deployment.

Flashlights and headlamps with rechargeable batteries are best especially if the same batteries can be used in all the appliances. Always good to have an extra battery that can be charged in the field. Newer LED flashlights can project a beam that can be seen for several hundred yards and put out over 1000 Lumens (Lumen is a measure of visible light from a light source)

A headlamp is a must have especially when working after dark. Some headlamps have the capability to project red light which is utilized to preserve night vision for viewing maps and operating equipment at night. Red Light flashlight are also available.

A Blacklight device will cause body fluids to glow. For example, blood spatter on carpet will light up under black light. It can also be useful checking fluorescence in Abduction and Mutilation Cases.

Laptop Computer

It goes without saying that a laptop is essential gear. A good Laptop, newer and faster is best. It should have several USB ports, a HTML port and SD card slot so SDXC Cards can be taken out of the camera and plugged into the computer.

Hi-Definition touch screen with lighted keyboard is recommended. It is nice to have a solid-state hard drive for faster start up and a longer battery life.

Most cellular smartphones have a mobile hot spot which allows the laptop to connect to the internet via the wireless cellular phone system digital modem, which is essential to check Google Earth Maps, access Star-Chart programs, check email and not to mention making hotel reservations.

User Manuals for Equipment

It is good to save a PDF (file) of the user manuals for each camera and other equipment you are using, on to your laptop hard drive. It is nice to be able to access the camera manual from your laptop if you are trying to set up a camera for a certain situation while you are on location.

Portable Hard Drive

Newer Solid state hard drives are available that are small waterproof and shock proof for use in the field (pictured here). It is best to upload all the SD cards from the cameras to the hard drive and editing is never done in the field.

Data-Security Thumb Drive

This is a high-level data security device used by the us government that has passed numerous tests relating to the encryption algorithm and tamper-evidence as well as to thwart brute force hacking. It is available in several capacities starting 4 gigabytes. It is waterproof and shock proof when placed in the aluminum O-ring sealed sleeve case (pictured here)

NATO
OTAN
Certified to NATO Restricted Level

The datAshur PRO uses military grade hardware encryption, which encrypts all data stored on the drive in real-time. It requires no software and incorporates a rechargeable battery allowing the user to enter a PIN (Personal Identification Number) onto the on-board keypad before connecting the drive to a USB port.

If the user enters an incorrect PIN 10 (ten) consecutive times, then both the encryption key and all data will be deleted and lost forever. Should the drive be lost or stolen, the user can rest assured that all data held on the datAshur PRO is safe and cannot be accessed by any unauthorized person.

Portable Power supply

This small portable power unit on the next page, produces 120-VAC for charging camera batteries and powering Laptops. Has USB ports for charging cellphones, tablets, and flashlights and a 12-volt receptacle for other 12-volt automotive accessories.

It weighs about 12 pounds and can be charged by the vehicle while traveling. It can also be charged by a solar panel for working off grid. This unit will supply power for a two-person crew with cameras and laptops for 3 days. It is excellent quality very well built and exceptionally durable.

Compass

Is a device that has magnetized a needle that aligns itself with the horizontal component of the earth's magnetic field. The needle points to the North magnetic pole. Some compasses have an" inclinometer" (elevation indicator) built into the unit as well.

Newer designs incorporate what is known as a Globally Balanced compass needle. This type of compass provides an accurate reading anywhere on earth with a rare earth magnet that resists interference and will not lose polarity.

The most obvious use of a compass is for navigation when using topographic maps. However, a handheld GPS would be the preferred navigation method referred to later in this chapter. A compass can be used to detect a magnetic field on a metallic object like a vehicle. Remember in science class putting a strong magnet on a nail, after a while the nail becomes magnetized. The same concept applies to a vehicle in a close encounter situation where a hovering UAP gets close to a vehicle.

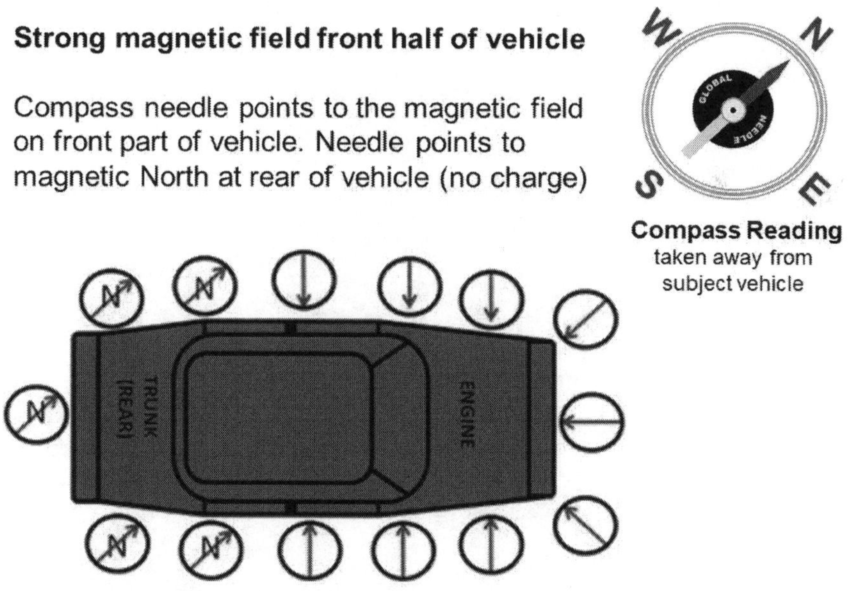

This illustration was done for a case I worked on. The readings were taken about 36 hours after the event. The field will eventually dissipate but will still be detectable for at least a month after the event. The anti-gravity propulsion (electrogravitic) emits a type of radiation, kind of like an electromagnetic field that is similar to microwave. This can actually magnetize a vehicle leaving a static field around the car. Typically, the field is stronger in the front where the most metal parts of the engine and transmission are located.

Radar Detector

A radar detector is a portable electronic device used by drivers to detect if their speed is being monitored by police and are used to reduce the chances of getting a speeding ticket.

Radar detectors are built around a superheterodyne receiver that can detect frequencies of electromagnetic waves in the radio or microwave frequency range.

I remember reading a couple of old case reports from back in the 1980s about close encounters that involved automobiles that had radar detectors inside (referred to as fuzzbusters). Apparently when in close proximity to hovering crafts the radar detectors sounded an alarm and one case the alarm went off and then the device burned out.

In the late 1990s I participated in a number of Night Watches. I owned a battery-operated radar detector and always set it up near the camera equipment pointed up. I never got an alarm but would be interesting valuable information.

In the event you are using a Multiple Camera Scenario (chapter 8), I would recommend placing the radar detector in view of the video camera set at wide-angle to view the entire area which should be running continuously.

GPS (global positioning system)

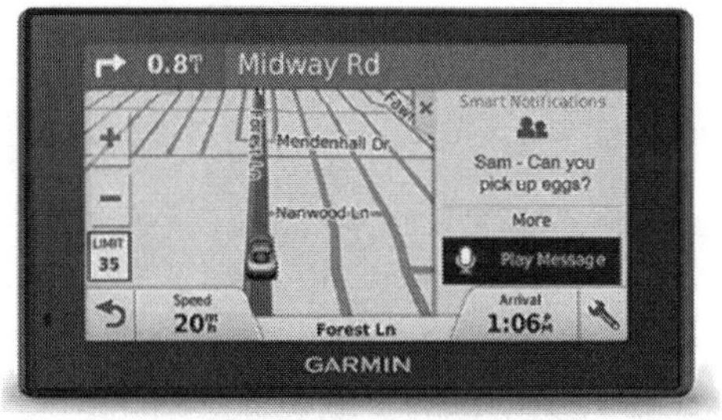

Automotive Navigation

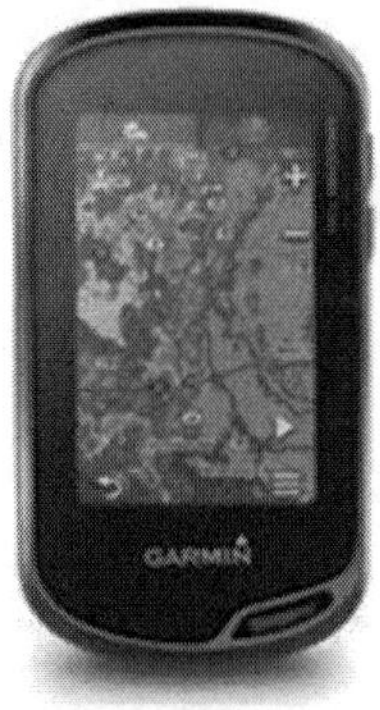

Handheld for off-road

GPS for highway navigation is necessary for locating witnesses and events. Always make sure the addresses for all the places you need to investigate are programed into the GPS prior to departing on an investigation.

Handheld units are nice for hiking off road and marking locations (of events) that may be away from roads, in forests and wooded areas. When leaving your vehicle, it is advisable to mark the location of the parking area so you can always find your way back to the vehicle even after dark.

Both Automotive and Handheld units can record GPS coordinates which are expressed as the combination of latitude and longitude. As mentioned in Chapter 5 most organizations now require GPS coordinates for UFO/ UAP Reports.

GPS Satellite Communicators

GPS with Satellite Communicators provide Global Connectivity via two-way messaging by connecting to a satellite network.

In areas where cellphone coverage is not available, text messages and email can be transferred via the Global Iridium® Satellite Network.

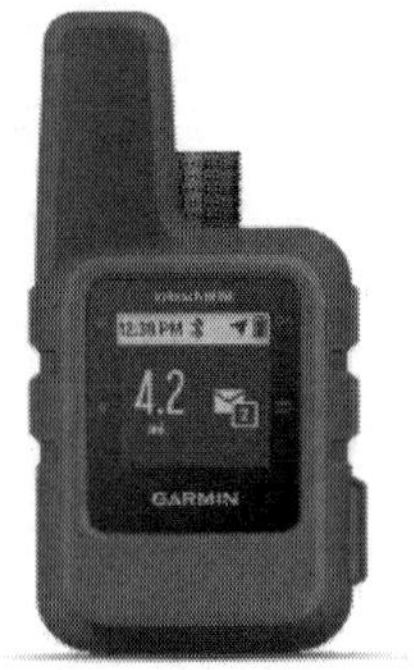

InReach Mini

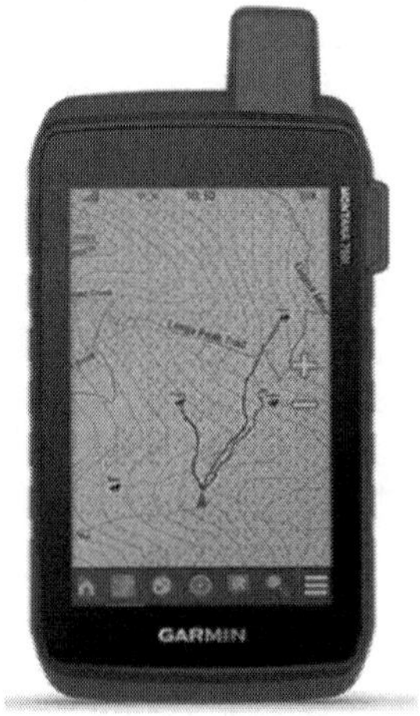

Montana 700i

A service plan is needed for Iridium® Satellite Network which usually cost about $12 to $25 per month depending on how many messages are sent during the month.

Email contacts and cell phone numbers can be stored onto the device. Email messages and texts can be typed on the touch screen and sent with coordinates of the location from where the message was sent. The email message will have a link to google maps. By clicking on the link, a map will open on the receiver's computer screen which will pinpoint the exact location of the sender.

During the BAASS SIP Project several of the investigators had GPS Satellite Communicators. While on a location of an incident they could send a message to indicate the exact location of an incident. For example, one of the investigators had a motor home, in the evening he would send me an email message transmitted from the Satellite Communication system with coordinates. When I opened the email and clicked on the link, you could see which parking space his motor home was parked in at the campground.

In case of emergency, an interactive SOS message can be sent to the Garmin International Emergency Response Coordination Center which provides professional 24/7 global monitoring and response center.

Walkie-Talkies

Walkie-Talkies are useful for communication with teams of two or more people.

Pictured here is a Garmin Rino GMRS two-way radio which transmits a full 5 Watts and has an extended range of about 20 miles.

It also includes GPS with a position reporting feature which shows the locations of the other users on the GPS map screen. As indicated, walkie talkies are useful in areas where there is no cellular service. Also, with several people on the team, any message that is transmitted goes out to all the other users simultaneously. (As opposed to calling 4 people on their cell phones). These units provide a great way for a team to communicate, while displaying the GPS location of each of the team members. These radios require an FCC license which is easy to obtain for a fee with a mail in application.

Note: the GPS equipment features herein is manufactured by GARMIN. I have no association with GARMIN other than having used their products for many years and found them to be extremely reliable and well made.

IPX8 military rating

An IPX8 rating means the device withstands incidental exposure to water of up to 1 meter deep for up to 30 minutes and is shock proof and dust proof.

The GPS, Walkie-talkies and Satellite Communicators featured herein have a IPX8 rating. I recommend that other equipment like flashlights, headlamps, cameras and battery power packs also be rated the same for use in the field.

Extension of the 5 Senses

Investigators use equipment to get sensory information from outside the human 5 physical senses.

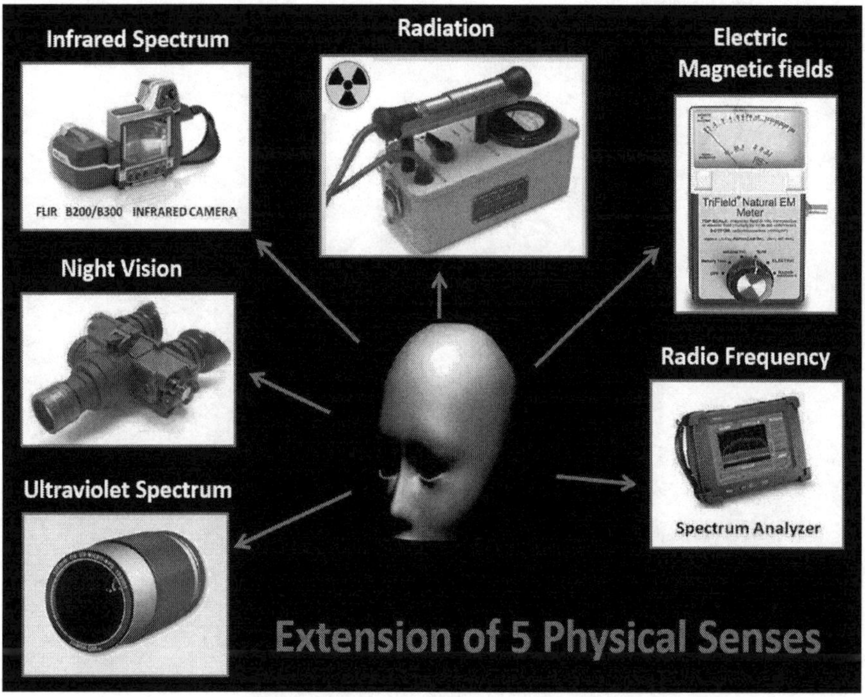

Light which is visible to our human eyes represents just a tiny percentage of the full spectrum of electromagnetic radiation, as illustrated on the next page.

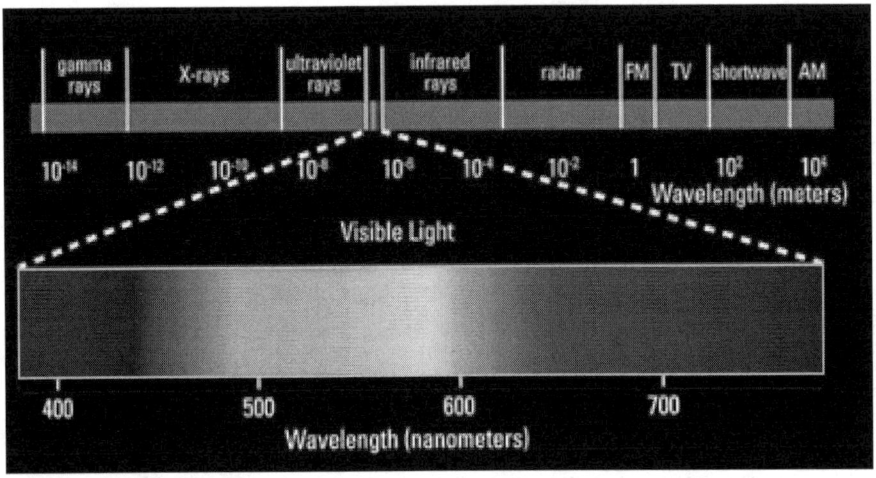

TRI-Field Meter

The TRI-Field Meter is Used for accurate measurements of electromagnetic fields (EMF).

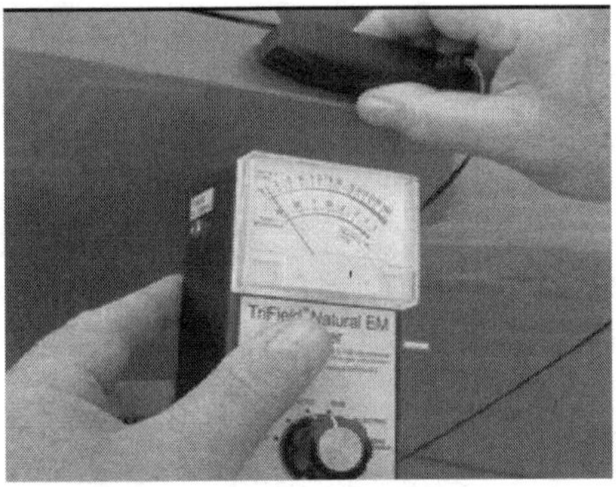

Pictured measuring the magnetic field on a vehicle.

This meter detects changes in natural and man-made electric and magnetic fields and is commonly used by paranormal investigators.

It is so sensitive it can detect you waving your hand near the meter or a person walking nearby, even through a wall. The sensitivity knob on the side allows you to adjust to the right amount you need for the environment. Just set it down, wait for the meter to baseline then turn to adjust the sensitivity knob to your preference.

Night vision

Night vision devices collect and intensify ambient light, starlight, moonlight and any other light in the night sky to allow you to see into the dark with vivid detail from extended viewing range in 100% Darkness.

Many night-vision optics are paired with an infrared light to help illuminate a field or target in the pitch-black darkness, especially on moonless nights.

Night vision goggles can be used to study the night sky and stars. An interesting experiment is to lay on your back and look up at the night sky using good high quality Night Vision. You will be surprised to see much more detail and objects that are not observable with the human eye.

Another point about night vision that I learned in flight school is that the light-sensitive nerves inside of the back of the eye, known as the retina, are more sensitive in the peripheral areas, than when looking straight forward. So, at night, if you have ever experienced seeing something out of the corner of your eye, then looking directly at it and it disappears, this is because your peripheral vision is more sensitive to light.

Thermal Imaging

Forward-Looking Infrared (FLIR) Unlike night vision, thermal optics read heat. The high-tech processors in a thermal optic camera detect the minute and large differences in heat and create a digital picture of the landscape in the optic viewing screen.

Handheld Thermal Imaging FLUR Scope

Helicopters used for Search & Rescue have Forward-Looking Infrared cameras which can pick up body heat from a lost victim. Law enforcement and firefighters use Thermal Imaging FLUR equipment to see inside of buildings to determine if individuals are inside.

Spectrum Analyzer

The primary use is to measure the power and the spectrum (frequency) of known and unknown radio signals. Looking at the radio frequency spectrum of a signal to examine whether any misleading or unwanted signals are present.

Pictured here is a handheld spectrum analyzer that can be used in the field during an event to identify radio signals. These instruments are awfully expensive, costing several thousand dollars and the opportunity to use them is very limited, so the average investigator would probably not have one available, as we did during the SIP Project.

Interestingly a handheld spectrum analyzer can be used to find electronic transmitter devices used in wire-tapping, commonly known as bugs. During the SIP Project several of us (including me) had our homes scanned looking for monitoring devices that may have been placed in our homes.

Ultraviolet Visible Spectroscopy

This Lens which is fitted with a Nikon mount will work on a 35mm camera or a DSLR. Very Hi-tec equipment that will cost over $6000.00, and It is designed to photograph an electrostatic field.

Electrical field TESLA COIL
Visible Range

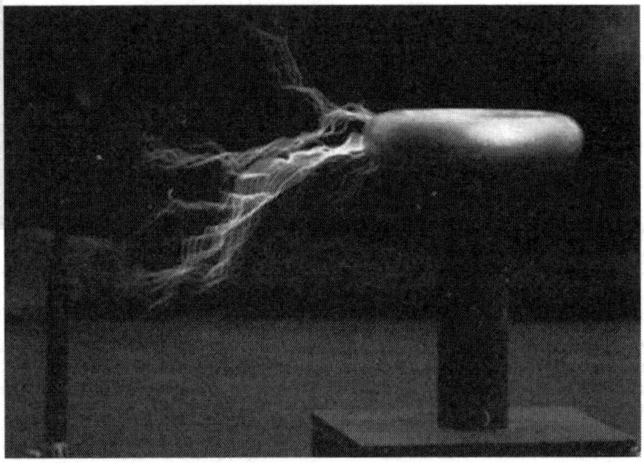

Image filtered for 254nm UV light

The image on the left is a photograph of a Tesla Coil taken in the visual range as seen with the human eye. The image on the right was taken with this Ultraviolet Visible Spectroscopy lens, set at 254 nanometers (wavelength) and shows the electrostatic field.

Once again, an expensive piece of equipment with limited use but a photograph set at 254 nanometers of an object at close proximity to the ground would be priceless and very useful in determining the fields emitted from the object.

MADAR

MADAR is an affordable and miniaturized version of a time-tested UFO detection system that went operational in May of 2018, now boasting 140 MADAR sites in the U.S. and around the world.

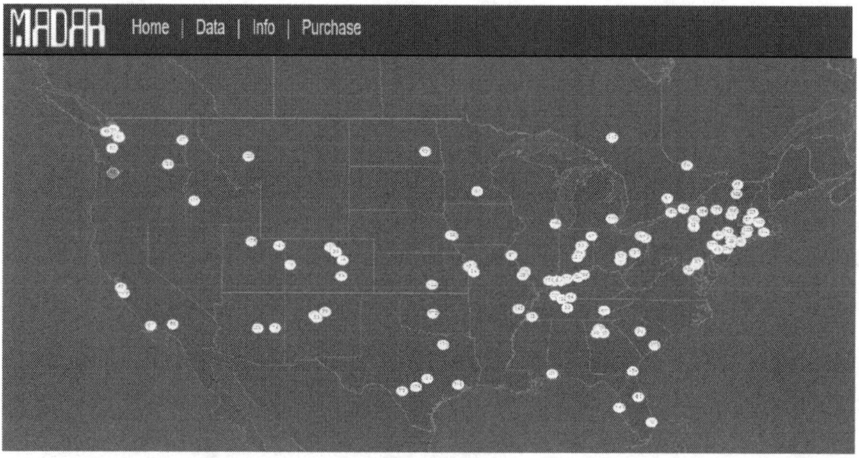

Live Map

Alerted operators can obtain even more useful data with handheld equipment, but the device already gathers data on local geomagnetic field, changes in compass heading, and data from an accelerometer, all time-stamped and stored on the MADAR server so that anyone can look at the data.

MADAR-III DATA PROBE
$240.00

DAS V2.0 ALARM
$45.00

UFOs are known to cause disturbances in the geomagnetic field and there are over 150 documented instances of compass-needle changes and UFO observations.

The device stores data every minute but ramps up to once every second in alert mode. The spreadsheet is available to anyone, but the alert notification goes out to the operator via cell phone, an alarm box, and documented in an email. The device has an onboard 1.2 GHz computer with 1 GB RAM.

You can join the MADAR Project by purchasing a MADAR-III Data Probe for $240.00 which includes the power adapter, controller, sensors, and software shipped Priority Mail. Ethernet cable not included.

The MADAR-III Data Probe comes preconfigured and is 100% PLUG AND PLAY. $240.00, includes 24/7 server. You can order here https://madar.site/

Flightradar24

Flightradar24 is a global flight tracking service that provides real-time information about thousands of aircraft around the world.

Flightradar24 tracks 180,000 flights, from 1,200 airlines, flying to or from 4,000 airports around the world in real time. https://www.flightradar24.com/37.96,-87.39/6

The service is currently available online and for your iOS or Android devices.

flightradar24
Apps
Add coverage
Data / History
Subscription plans
Social
Press
About
Commercial services
5. MOQSE34
Statistics
Terrestrial ADS-B 1,471 12,107
Satellite ADS-B 5 194
MLAT 8
Radar 5
OGN/FLARM 0
Estimated 11 218
TOTAL 1,500 13,480
Tweets
Blog posts
App Store
Google Play
United States
Settings
Weather
Filters
Bookmarks
Widgets
Playback

END CHAPTER

Chapter 9

Perception of Reality

During the early stages of our lives as a result of our education, family situation and our environment, we all develop a perception of reality; what we believe is possible (or not possible) and how the basic laws of physics affect our lives and the way we live.

Studies indicate that by age seven we actually learn almost half of everything we will learn in our life. Think about it, shortly after we are born, our vision and sensory perceptions develop. We learn to see, feel, taste, smell, touch and feel pain. By age two, we begin to talk and walk. Then we begin to interpret facial expressions, learn how to express emotions, and ultimately communicate with those in our environment.

Next of course we learn the rules. The rules are typically made by the people who raise us, care for us, and control us. By age seven we are fairly well locked in by the ideological concepts of our parents, our schools, religious organizations, government agencies and peer pressure of those around us.

Most important as we get a little older, we begin to understand the constraints of our three-dimensional physical reality. By constraints I mean physical space (length-width-depth) and most importantly time. Psychic research centers and most religious organizations will ask the prospectives to acknowledge that they are more than their physical body.

The concept of consciousness may be how we define and understand our own existence within our three-dimensional physical reality and the non-physical part of our existence.

Regardless of where you were born or how you were raised, you have been programmed. You have conscious presence of mind and understand who you are, were you live, what you do, how much you are worth, and who your friends, family and enemies are.

But on a subconscious level you have been subject to immensely powerful programming, in terms of your belief systems and what you think is possible (or not possible).

We learned early on that fire can be dangerous, as a child reaching out to touch the flame of the candle, learned quickly that fire will burn human skin and cause great pain.

In contrast I recently watched a videotape of an individual I know who was attending a seminar on mind control that involved intense classes and meditation. During the experiment, the facilitators scraped together a pile of red-hot coals and spread them out on the ground. The instructor took a sheet of paper and dropped it over the pile of red-hot coals, it burst into flames about 10 inches above the surface of the hot coals from the intense heat, then the individual walked across the bed of hot coals in bare feet without sustaining any burns or injuries.

Monroe Institute.

There is a research center located in the mountains south of Charlottesville, Virginia, known as the Monroe Institute. The Monroe Institute is world renowned, known for its research in the study and exploration of human consciousness. I have been there a couple of times for short weekend classes and know many people who have attended classes there over the years. One of my long-term goals is to attend some of the more intense classes on out-of-body experiences (Gateway Voyage). I would highly recommend anyone seriously trying to understand the UFO phenomena look into classes there.

It is located in a beautiful mountain setting at Farber, Virginia. The campus consists of numerous buildings that house living quarters, meeting rooms, sleep chambers and dining facilities. I remember a place like a living room, where a framed glass case hangs on the wall. Inside the case are a number of metal forks and spoons that are all bent and twisted up. They looked like they were melted then twisted up. Pictured below taken with my cell phone.

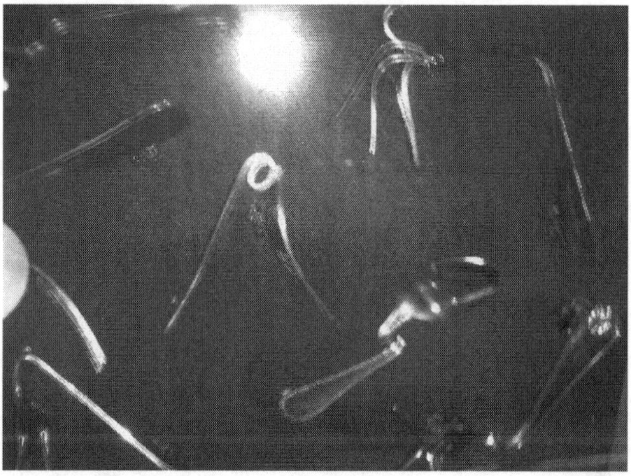

Interestingly, during my time working in the airports in central Virginia for the U.S. Department of Homeland Security, I was aware that the TSA screeners at Charlottesville airport often found bent up spoons and forks in traveler's luggage, typically people who were returning home from classes at the Monroe institute.

On several occasions while working, I happened to be there while their luggage was being screened and the twisted utensils were being found on the x-ray. I would smile saying "I see you have been to the Monroe institute", I'm sure they were initially shocked, wondering how the Feds knew they were at the Monroe institute, but once I explained that I knew a little about research at the Institute, I was able to hear some exciting stories while they were waiting for their flights. Oh yes, the travelers were ultimately allowed to take their souvenirs home with them because at that time forks and spoons were not prohibited items. I remember how excited and energized these people were after having taken classes at the Monroe Institute.

I know several people who attended classes which were in part to learn how to focus mental energy and actually bend a spoon as described. A mature adult in our modern society would have an extremely difficult time learning how to do this, the reason being that we all know that this is impossible, based on our subconscious programming.

I am quite sure that it can be done because I have seen the evidence firsthand. However, an individual would have to focus a tremendous amount of energy to change their perception of reality. Research indicates that children four or five years old, seem to be able to learn easily, because by that age they're not really aware how impossible it is.

Turning the Corner

I would be the first to admit that during the early days of my exploration I saw things and worked on cases that I did not understand completely. My scientific education, Bachelor's degree in Aeronautics and my training as a commercial pilot were obviously formative in my thinking.

My perception of reality has drastically changed since that flight in 1976 and I look at case report data much differently now. Partly because I have had the chance to analyze hundreds of witness case reports and interview many people associated with abductions and close encounters over the course of my research. A couple of examples of how my perception changed may help illustrate my point.

Inter-Dimensional Portals

As indicated in 1975 I was a flight student in Florida. During a routine flight training exercise, I was flying a single engine Cessna 172 in the midafternoon under a beautiful clear blue summer sky over the everglades. I was alone in the aircraft at about 2500 feet and noticed a disturbing vision which was at high altitude and great distance in the northern sky. I saw what appeared to be a rectangle shaped window or portal. It was a beautiful Florida summer blue sky and there was this opening which looked like a frame with fire burning in it. It looked like an orange glowing flickering effect in a rectangle shaped portal in the sky. Holding a credit card at arm's length would give one a good perspective of the size and shape it was in my view. It almost seemed impossible, but it was definitely there.

I observed it for over an hour as I finished my training maneuvers, noting that it remained stationary in the sky. I returned to my airport and told the flight dispatchers at the flight school of my observation. It was conveyed to me that this was not anything to worry about, as the Airforce had dropped tin foil to test radar in the area and it was sunlight reflected from the foil that created the orange glowing effect I had witnessed. I totally believed their cover story and was actually relieved.

Many years later I was privy to several credible case reports about witness observations of what is believed to be inter-dimensional portals and realized what was described in the reports was exactly what I had seen during my training flight. My perception of reality has obviously changed, and I can look back on that event with a different understanding based on my experience and research. We will talk more about Inter-Dimensional Portals in Chapter 10.

Devil's Triangle

I had another very strange experience in Florida that same year, when flying out over the ocean in the area referred to as Devil's Triangle, (also known as the Bermuda Triangle).

The region is loosely defined as a triangular area measured from Florida, southeast to Puerto Rico and north to

Bermuda, that covers about 500,000 square miles of ocean in the North Atlantic Ocean.

This area is a mysterious place where many ships, boats and aircrafts have disappeared under inexplicable circumstances and never been heard from again.

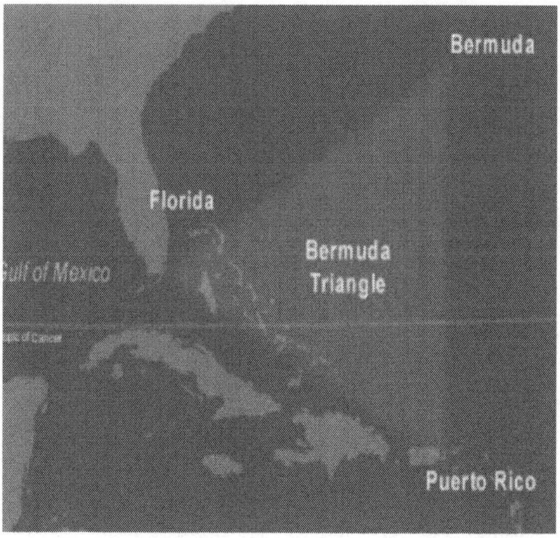

Many believe Opa Locka airport is the tip of the Devil's Triangle, it is located just north of Miami, Florida, and was a U.S. Navy Military Airport during WW II.

Opa Locka airport has the distinction of being the airport where more aircraft have taken off from and disappeared (never returned) than anywhere else on Earth, including Amelia Earhart who took off from there on her ill-fated flight around the World in 1937.

I spent some time at Opa Locka Airport in 1975 completing flight training. Back then it was more like a deserted military base with a flight school and several small general aviation businesses. Today it is a huge bustling remarkably active municipal airport.

In any event on a beautiful sunny day, I departed from Opa Locka flying a single engine aircraft, while doing some flight training at about 2500 ft AGL. I flew out over the ocean, headed Southwest for about 60 miles when I spotted a large ship on the surface.

This started out as a routine training flight but after flying over the ship, I noticed some of my flight instruments were appearing to malfunction. There was a magnetic compass mounted in the windshield that started to wabble and rotate. At the same time, my directional gyro compass, which is a mechanical instrument with a compass dial that is gyro operated, started to slowly spin. The communication radios and the navigation radios appeared to malfunction going dead as well.

This was all very weird and a little scary but not life threating at that point, because the weather was clear, and visibility was excellent. Also, the ship was still there to help me stay oriented (by direction). I was understandably nervous and decided to fly back to Opa Locka, which I did without incident. When I returned (close to the Florida coast) everything in the aircraft seemed to be operating normally.

Point being that if the weather and visibility would have been bad flying in instrument conditions or at night, the outcome would have been perilous if not fatal. There was no explanation for this malfunction only that there is some strange unnatural weird force out there, which I have experienced first-hand. It obviously helps to explain why so many ships, boats and aircrafts have disappeared.

Apollo 11
Encounters with Extraterrestrials on the Moon.

This is an amazing story that will push most people to the edge of their perceptual comfort level. It is about close encounters with extraterrestrials during the Apollo 11 Space Mission and first walk on the Moon. I am quite sure this is true and accurately relayed herein, as I have researched it thoroughly. However, I am convinced it would challenge most people's perception of reality.

The radio communication between NASA Houston Space Center and the Apollo 11 crew was intercepted by numerous amateur HAM radio operators on earth. This was obviously a time before sophisticated radio encryption equipment, which would be used today that would prevent unauthorized radio operators form receiving such transmissions.

ECHO Magazine Article

Several years ago, I published an article in a local magazine about this Apollo 11 event. I spoke with someone I knew well in Charlottesville telling him about the article that I was going to publish in the next edition,

Pictured on the next page is the graphic from the Magazine article I published with ECHO Magazine.

My friend told me about a special project he had worked on at the McCormick Observatory, in early months of 1969. This individual is someone I know to be highly credible, well-educated to the level of PHD and has asked to remain anonymous.

McCormick Observatory, Charlottesville Virginia

At that time, the telescope at the McCormick Observatory (pictured on the previous page) was state of the art, highly sophisticated and one of the two most powerful telescopes in the world. The purpose of his project was to explore the surface of the Moon in order to identify potential landing spots for the Apollo 11 Space Mission. He articulated to me that while exploring the surface of the moon with the telescope, he personally viewed UFO traffic and lights on and over the surface of the moon, just prior to the Apollo 11 Space Mission.

In 1979 Maurice Chatelain was the Chief of the NASA Communications Systems. He made a very public statement that: "All Apollo and Gemini flights were followed, both at a distance and sometimes also quite closely, by space vehicles of extraterrestrial origin. Every time it occurred, the astronauts informed Mission Control, who then ordered absolute silence."

Gordon Cooper, a former USAF Pilot and one of America's original seven Mercury Astronauts famous for his Mercury 7 space flight acknowledged that he personally saw and filmed UFOs. He was outspoken about the need for an open inquiry into UFO enigma. In 1978, he sent an open letter to the United Nations in which he stated: "I believe that these extraterrestrial vehicles and their crews are visiting this planet from other planets, which are a little more technologically advanced than we are".

Walter Schirra, while on the Mercury 8 space flight, was the first of the astronauts to use the code name "Santa Claus" to indicate the presence of flying saucers next to his space capsule. According to Mr. Chatelain, Schirra's announcements were barely noticed by the "general public".

Mr. Chatelain also acknowledged (in his public statement in 1979) that: "Neil Armstrong had indeed reported seeing two UFOs on the rim of a crater. The encounter was common knowledge in NASA, but nobody has talked about it publicly until now."

Apollo 11

Apollo 11 was the first space mission to land on the Moon. Neil Armstrong, Michael Collins, and Buzz Aldrin manned this lunar flight mission. Collins piloted the command module around the Moon (remaining in orbit) while Armstrong and Aldrin descended in the Lunar Landing Module (LEM), touching down on the surface of the Moon at 4:17 P.M. at a location on the Moon known as "Sea of Tranquility" on July 20, 1969.

Neil Armstrong relayed messages to Houston Mission Control that two large, mysterious objects were watching them after landing near the lunar landing module. But his message was never heard by the public because NASA censored it.

Christopher Kraft was director of NASA Houston Space Center during the Apollo projects. He made some interesting comments about this Apollo 11 event when he left employment at NASA. His comments about the transcribed excerpts from radio transmissions have been corroborated by hundreds of amateur radio operators who had tuned in their stations to the same frequency that the astronauts communicated with NASA Houston Space Center (Houston Control).

Ham radio operators who had their own VHF receiving facilities picked up the following amazing conversations. At the same time the live television broadcast was interrupted for several minutes due to a supposed "overheated camera", but the transcribed radio transmissions (below) were received loud and clear by over 700 ham radio operators in the USA, while an estimated 600 million people around the world (myself included) were watching and listening to the official version on TV.

Buzz Aldrin and Neil Armstrong were walking around on the moon in view of the camera some distance from the Lunar Landing Module (LEM) when Armstrong suddenly clutched Aldrin's arm excitedly and exclaimed (as the camera and sound went off during the public live TV broadcast):

- **Apollo 11 (Moon):** (Neil Armstrong) What was it? What the hell was it? That's all I want to know!"
- **Mission Control: (**Christopher Kraft-Huston Space Center) what's there? ... (garble) ... Mission Control calling Apollo 11.
- **Apollo 11 (Moon):** These babies were huge, sir! ... Enormous! ... Oh, God! You wouldn't believe it! I'm telling you there are other spacecraft out there ... lined up on the far side of the crater edge! They're on the Moon watching us!

- **Apollo 11 (Moon):** Those are giant things. No, no, no this is not an optical illusion. No one is going to believe this!
- **Mission Control:** What ... What ... What? What the hell is happening? What's wrong with you?
- **Apollo 11 (Moon):** They're here under the surface.
- **Mission Control:** What's there? (muffled noise) transmission interrupted . . . Mission control calling Apollo 11.
- **Apollo 11 (Moon):** We saw some visitors. They were here for a while, observing the instruments.
- **Mission Control:** Repeat your last information!
- **Apollo 11 (Moon):** I say that there were other spaceships. They're lined up in the other side of the crater!
- **Mission Control:** Repeat, repeat!
- **Apollo 11 (Moon):** My hands are shaking so badly I can't do anything . . . Film it ... God, if these damned cameras have picked up anything . . . what then?
- **Mission Control:** Have you picked up anything?
- **Apollo 11 (Moon):** Three shots of the saucers or whatever they were that were ruining the film.
- **Mission Control:** Control here . . . are you on your way? What is the uproar with the UFOs over?
- **Apollo 11 (Moon):** They've landed here. There they are and they're watching us.
- **Mission Control:** The mirrors, the mirrors - have you set them up?
- **Apollo 11 (Moon):** Yes, they're in the right place. But whoever made those spaceships surely can come tomorrow and remove them. Over and out.

During a NASA symposium several years later a professor who wished to remain anonymous was interviewing Neil Armstrong about this event. Note: Buzz Aldrin apparently took color movie film of the UFOs from inside the module and continued filming them after he and Armstrong went outside.

- **Professor:** What REALLY happened out there with Apollo 11?
- **Armstrong:** It was incredible, of course we had always known there was a possibility, the fact is, we were warned off! There was never any question then of a space station or a moon city.
- **Professor:** How do you mean "warned off"?
- **Armstrong:** I can't go into details, except to say that their ships were far superior to ours both in size and technology. Boy were they big! And menacing! No, there is no question of a space station.
- **Professor:** But NASA had other missions after Apollo 11?
- **Armstrong:** Naturally, NASA was committed at that time, and couldn't risk panic on Earth.

Armstrong confirmed that the story was true but refused to go into further detail, beyond admitting that the CIA was behind the cover-up."

In a published article in April 1995, Maurice Chatelain who was former NASA Director of Communications, dropped the bombshell revelation stating that, "the Apollo Moon Missions found 'several mysterious geometric structures of unnatural origin' on the Moon".

SECRETS OF THE APOLLO MOON MISSIONS

During the Apollo moon missions, some of the astronauts participated in psychic experiments associated with ESP (Extrasensory Perception). My physics professors at Embry Riddle, had worked on the Apollo projects at NASA, and told me about ESP experiments where astronauts trained in identifying symbols on cards with individuals that they were able to develop a psychic connection after much practice.

Basically, two people sit across from each other, one person looks at symbols on a card, which is drawn from a deck, and the other person tries to visualize the symbol. The symbols were simple circles, squares, check marks, plus and minus signs, some may have had numbers. The team would sit face to face across a table that was divided by a board so the receiver, who would try to visualize the symbol, could not see the cards on the other side of the table.

There were several astronauts who were able to develop relationships with ground personal where they had established an extremely high accuracy rate between them. The idea was to conduct the tests when one of the astronauts was in orbit around the moon and the other person was on the ground at the space center, in order to understand the effect that distance would have on the accuracy. As planned in a very carefully timed experiment, the astronaut in orbit of the moon would open a stack of cards and view each of them at exact times which were recorded. The experiments were timed meticulously.

Two things were learned; first that the distance had no effect on accuracy. The accuracy rate was exactly the same as when they were sitting across the table from each other.

The second and unexpected result was that the distance had no effect on the timing. It should be noted that the distance between the earth and moon causes a delay of about 2.5 seconds in radio transmissions. It takes about 2.5 seconds for the radio signal to travel form the moon to the earth. You will remember watching the Apollo space flights on TV, as we listened to the astronauts talking to Houston control center, hearing voice transmission and then a bleep sound. The bleep is an end of transmission signal that occurs when the speaker releases the transmit button on the microphone.

Psychic Communications

Light travels in a set direction and at a defined speed (186,000 miles per second) radio waves travel at about the same speed as light, and hence the delay of several seconds between the conversation and the corresponding bleep signals.

The important thing to understand here is that psychic transmissions do not conform to the constraints of our time space reality. Telepathy (transmission of information from one person to another without using any known human sensory channels or physical interaction) is timeless and instantaneous, regardless of the distance between the two entities in communication.

Distorted Reports

Unfortunately, much of what is reported involves the other aspects of UFO cases, particularly high strangeness which seems to disregard our current laws of physics (anti-physical) and the psychic effects that are so often reported by witnesses during these encounters, which will be discussed in the next chapter.

Sometimes they are ignored or reported in a very distorted manor, because investigators tend to listen to the witness and then write the report based on their own personal perceptions about reality and understanding of physics (as opposed to accurately recording the witnesses experience).
In chapter 2, I shared a case report about an incident that occurred in a now closed base known as Vent Hill Farms. The witness stated that: "*the beings came through the fence (metal chain link) and stood on each side of her with one behind her. The three beings took her back through the fence to the still hovering object*" so when I wrote the report,
I could have said, "the beings climbed over the fence", or that "the beings crawled under the fence, or that "the beings cut a hole in the fence" but in fact I reported it exactly the way all witnesses articulated it to me.

In chapter 10 Multi-Dimensional Consciousness we will look at how at higher levels of consciousness the density decreases and might help explain how someone could go through a metal chain link.

Take for example a case report where a witness makes a statement like this. "We noticed the object bright in the night sky, at high altitude and at a great distance, it appeared to react to my thoughts" or "its maneuvers somehow appeared to anticipate my thoughts as I became cognizant of it, suddenly it moved directly toward me and instantly stopped almost right over my head".

Looking back at such data in the culmination of many other similar occurrences, the consensus of thinking becomes much different. Take the same case in point with testimony like "it appeared to anticipate my thoughts," or "it appeared to react to my thoughts" these same words have been found in hundreds of witness sighting reports. The exact words are articulated over and over again in many reports, which says something about the event and the possibility that these witnesses may have been stating exactly what actually happened.

Consciousness

In physics class back in the 1970's we were taught that space, at least from an interplanetary perspective was a vacuum. Travel outside of the atmosphere of the Earth, space was simply empty of any substance and void. What we know now is that the universe is connected by a fabric or energy that occupies all space, which is consciousness. Space is not void as originally thought. Weather you study the ancient teachers, or quantum physics the understanding is surprising similar.

The real power is in the subconscious mind, which will be explored in more detail in the next chapter. For the record most of what is written herein about human consciousness comes from my years of research, essentially being in the trenches looking at evidence, interviewing witnesses and experiencers and working with other investigators.

Consciousness is the missing element in exploring the UFO/UAP phenomena. In our effort to understand the phenomena, we study the types of vehicles, try to understand the inter-dimensional method of travel and the energy sources used to power these incredible crafts. Trying specifically to understand who these entities are where they come from and why they come here.

Leave Your Perception of Reality at the Door

I attended a paranormal conference several years ago. As I was about to walk into the auditorium for one of the presentations, I noticed a big sign at the door in bold letters that said, **"leave your perception of reality at the door"**. Looking back over the years of my work I think this is probably pretty good advice.

We all have a well-established framework for our thinking as indicated, regardless of where you were born or how you were raised, your programming will affect your observations and perceptions, as well as how you communicate with other people.

During the course of my years of research especially while conducting an investigation or interviewing a witness, I normally would never tell someone what I believed or what I was thinking.

Leave your perception of reality at the door is good advice for any investigator to undertake in embarking on this new endeavor.

END CHAPTER

Chapter 10
Multi- Dimensional Consciousness

The slides and illustrations used in this chapter were developed from significant research I had done while trying to understand the different dimensions of existence and human consciousness. I developed a PowerPoint deck that I have since used many times for my lectures and speaking presentations.

The first time I used these slides was at a UFO conference in Pittsburgh. I was extremely nervous about presenting this information and honestly expected to be challenged and ridiculed when I got to this part of my presentation.

To my surprise, I received significant positive feedback about this part of my presentation. Over the last several years I have used these slides in a number of presentations both at UFO conferences and at the National Paranormal Coalition conference where I was a key speaker. Each time the attendees came up to me after I finished speaking and told me they thought this was the most interesting significant part of the presentation. So put your seatbelt on and turn the page.

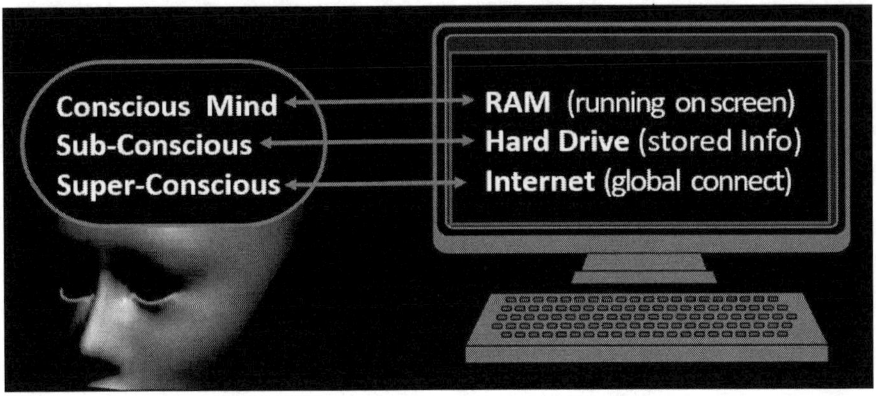

During my presentations and lectures I used this example to help illustrate the basic concepts of consciousness by comparing the human thought experience to the workings of a modern computer.

Conscious Mind

The Conscious Mind can be compared to what is running on the computer screen, which is known as the RAM memory. Active computer programing that is being played out is seen as images, graphics, video and heard as sounds, voices, and music on the computer. All calculations and processing are done here.

In the mind, this is the realm we think of as physical and present time, what is going on around us at any given moment. What we are experiencing now, cognitive thought involving the processes of thinking and reasoning as information comes into our brain through the 5 physical senses of sight, hearing, touch, taste, and smell. We know this as 3rd dimensional consciousness, which we will explore in the next few more pages.

Sub-Conscious Mind

The Sub-Conscious Mind can be compared to the Hard-Drive of the computer where everything that was ever viewed, downloaded, sent out, processed, calculated, and played is stored.

The Sub-Conscious mind contains the summary of all stored experience as memory. The brain records everything experienced in the conscious realm (from the 5 senses) in the Subconscious.

I worked with a regression hypnosis doctor (Harvard psychologist) while we were filming the Discovery Channel shows mentioned in Chapter 2. He told us that he could hypnotize someone and regress them back to something that happened to them in the second grade. He said I can get this person (in a regressed state) to look around the classroom and tell me what color shirt each classmate was wearing that day.

Think about it, everything you have ever seen, heard, felt, tasted, and smelled is recorded in your subconscious. When we remember we are simply calling up some of those stored experiences from the subconscious memory into the conscious realm of thinking. Kind of like opening a file on the computer.

It also works in the reverse; when you smell a certain fragrance (perfume) you are instantly reeled back 20 years to a romantic encounter, or you might hear music that takes you back to a wonderful wedding event that you enjoyed, and the sound of a police siren might place you in the middle of an accident scene memory.

Unfortunately, unlike the computer where a file can be deleted from the hard drive, the human subconscious retains everything; both good, bad, and fearful. In the previous chapter we talked about the programming that we had inflicted on us as we were growing up. Our belief systems are deeply rooted in the subconscious. The subconscious is where all the power is in terms of our perception of reality. The mind or the brain has access to the world outside only through the 5 senses. All reality or perception of reality is established in the subconsciousness.

Super Conscious Mind

The Super Conscious Mind can be compared to the way the computer connects to the internet for information that comes from the outside world. We know that the brain emits electromagnetic energy. An electroencephalogram (EEG) is a test that detects electrical activity in the brain. Research currently indicates that the heart also emits a type of electromagnetic energy that we now know is much more powerful than the brain as it radiates outward. If the brain is the center for conscious thinking, then the heart may actually be the center for emotional feelings.

Most of us believe that there is more to our existence than our physical body. We explore that by examining the connection to our higher self, as some refer to as the soul or spiritual guides. When faced with critical decisions sometimes people will say “sleep on it” because somehow, we connect with that higher self and wake from sleep with renewed creative energy and answers to our problems.

The heart somehow connects with those higher energy sources, which are expressed in the form of emotional responses. Metaphysical people call this the "emotional guidance system" because voices obviously don't whisper in your ear, but the communication comes in through the sense of emotional feelings, which are actually direct modes of communication. Did you ever get involved with something and immediately get a warm and fuzzy feeling about it? Could be your higher self telling you that this is good for you, but by the same token while being introduced to something new, you get a sick at your stomach feeling about it, maybe because you are being warned away from it.

Victims of violent crimes will often tell police investigators that they got a bad feeling right before being attacked. A co-worker of mine was about to leave a theater, via the back door that leads directly to the parking lot. As she approached the exit door, she got an unbelievably bad feeling that there was danger out there, and her better judgement was telling her to wait and go out the front door with the other people going to their cars. She ignored it, proceeded out the back exit, while walking to her car she was attacked, and her handbag was taken from her.

Dimensions

Dimensions are different planes of existence defined by a Vibratory Rate. Each dimension has defined parameters of existence and constraints that are in tune with the frequency of that dimension.

It is like when you turn on a television and select a channel, set at a certain frequency. You can watch a program and then change the channel by setting it to a different frequency. Both channels come in clearly when the frequency is set by the channel selector. Both channels exist simultaneously but you can only experience the one which is set to the frequency you are tuned into and watching. The Dimensions all exist simultaneously but the observer can only experience the plane of existence that is in tune with their frequency.

Consciousness

Consciousness is the awareness or ability to experience and function clearly, within that dimension, because the vibration is in resonance with the frequency of that dimension. (tuned into that frequency)

Unconscious

Unconscious is not functioning or being able to perceive stimuli within a given dimension. (TV is turned off or Individual is asleep). You can be unconscious in one dimensional reality perhaps while asleep but conscious in another dimensional like the dream state.

Multi- Dimensional Consciousness

Ancient Teachings and recent Quantum Physics both indicate and define the existence of 12 dimensions. There is much written about this research in the world of Quantum

Physics and the search for other dimensions, but for the purposes of this book, I want to explore only the first 5 dimensions. Keep your seat belt on and stay with me for another few pages, hopefully it will all make more sense.

1st Dimensional Consciousness - MOLECULAR
2nd Dimensional Consciousness - BIOLOGICAL
3rd Dimensional Consciousness - PHYSICAL REALITY
4th Dimensional Consciousness - ASTRAL DIMENSION
5th Dimensional Consciousness - SOUL AND SPIRIT

1st Dimensional Consciousness
MOLECULAR

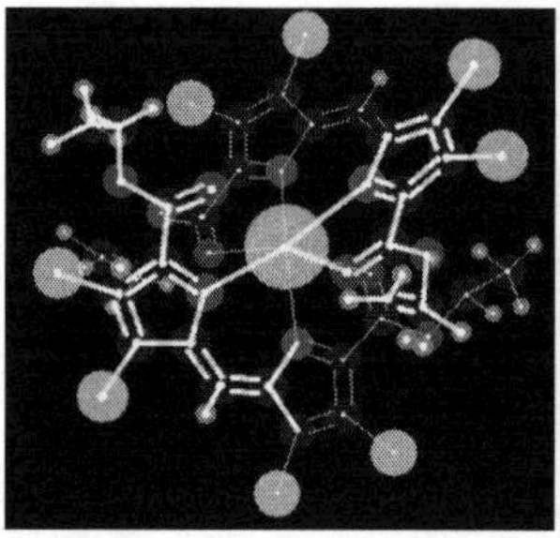

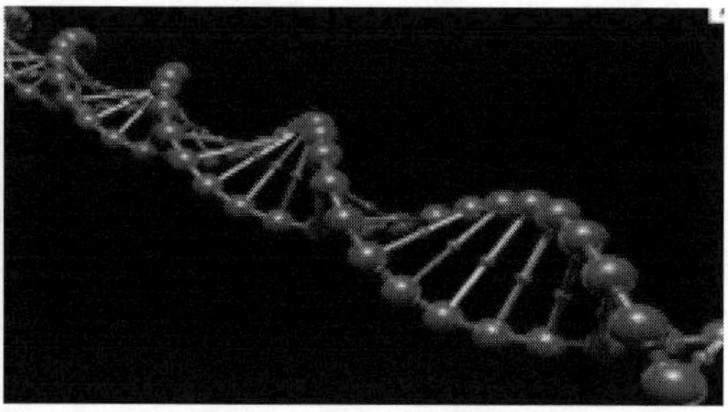

This is Foundation of our Physical Forms, in the state of the microscopic atomic structure of mater. A world of electrons protons and cellular nuclei, as well as our genetic codes in our DNA strands.

We understand the molecular structure of these elements through scientific research, recent developments in quantum physics and the use of electron microscopes, but obviously we cannot perceive any of this with our physical senses, so we are unconscious in this molecular dimension, although we are connected to it physically and we believe it all exists.

2nd Dimensional Consciousness
BIOLOGICAL

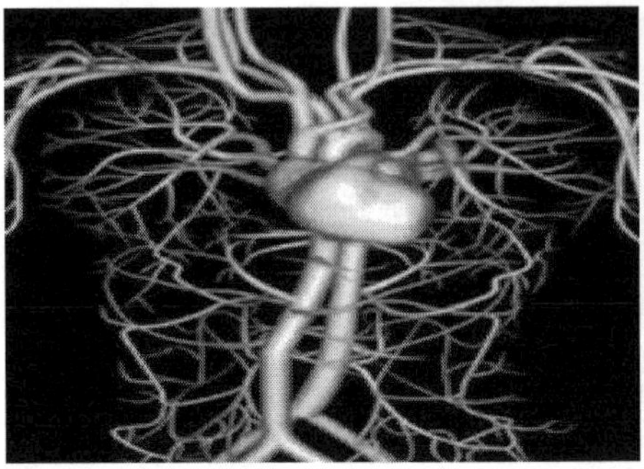

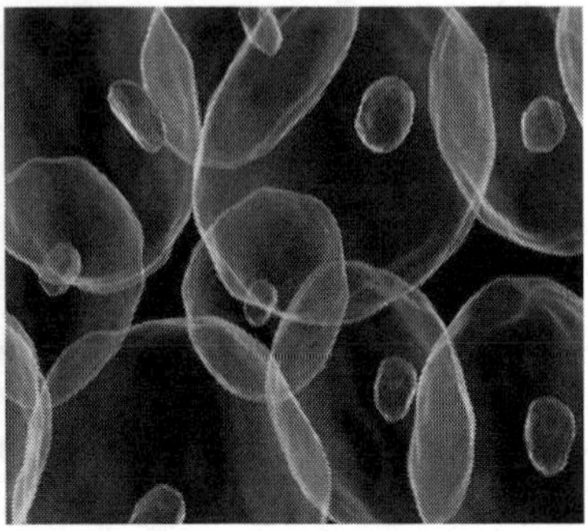

The Autonomic Nervous System centered in the lower part of the brain regulates all life support functions and the communication networks for the cellular organisms within the body. We understand how this works but do not have any conscious connection to the Nervous System. You do not have to think to make your heart pump, so we are unconscious in the biological Dimension as well.

3rd Dimensional Consciousness
PHYSICAL REALITY

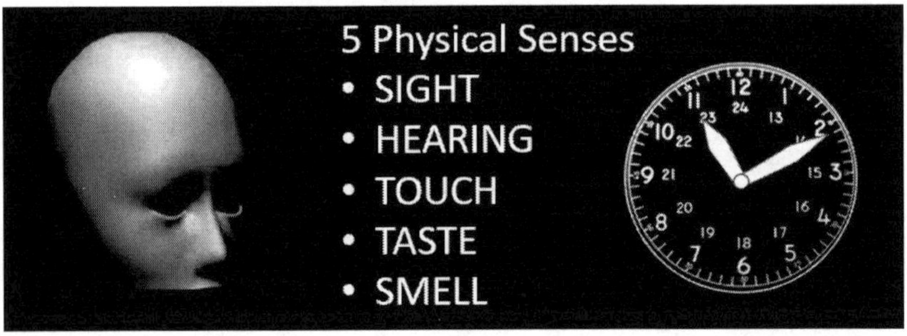

This is the world we live in as we use our 5 physical senses to experience everything in the conscious realm around us. It is a specific current time space reality that is defined by a specific frequency or vibration that we are in resonance with (in tune with) when conscious in this dimension.

In this dimensional existence, Linear Time is a constraint in the sense that we are subject to the movement of time, minute by minute, hour by hour, we cannot escape the swing of the pendulum or ticking of the clock. We can only experience the current moment. We remember past events but cannot go back in time and physically experience them again. We are locked in and constantly moving forward in time while not seeing what is beyond the next minute.

Physical Space is also a constraint defined by our surroundings in terms of length, width, and height. The ground we stand on or the room we sit in. It is also a world of cause and effect. If you fall you sustain an injury.

4th Dimensional Consciousness
ASTRAL DIMENSION – DREAM STATE

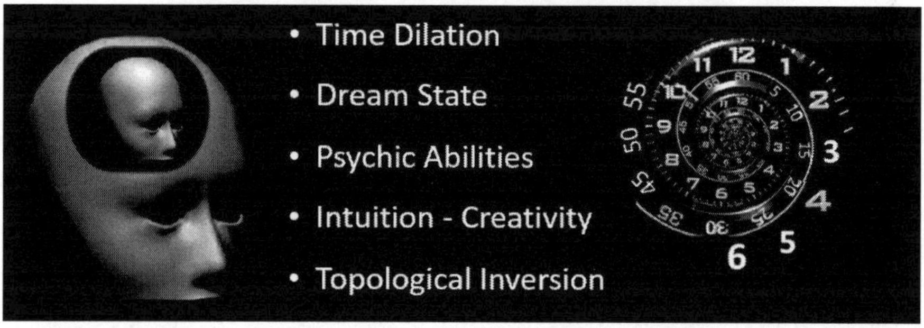

As the frequency increases in the 4th Dimension, we find ourselves in the bridge between the 3rd and 5th Dimension. In this state the 5 senses are not connected while asleep because we are unconscious in the third dimension but conscious in the higher frequency Dream State or Astral Dimension.

This is a realm where the observer can experience the past, the present, or the future, because the perception of time is not linear or constraining in this dream state a concept oftern referred to as Time Dilation.

Similarly the constraints of Physical Space in terms of length, width, height as well as cause and effect do not apply, in some cases producing what is known as topological inversion or space dilation. An example would be where the observer precieved an object that was relatively small on the exterior but huge in size on the interior.

Additionally what is explained as mobility of form, where the visual image that is observed can change to a different form. The image of a human can shift into the image of a different human or an entity that may frighten the observer, known as Shape Shifters.

This is also a realm where creativity originates from the superconscious connection and plays out for the observer. Psychic abilities and Intuition also originate in this state.

Polarization of Light and Dark can be understood and contrasted. The observer is preceiving what is good or bad, right or wrong, evil and divine.

5th Dimensional Consciousness
SOUL AND SPIRIT

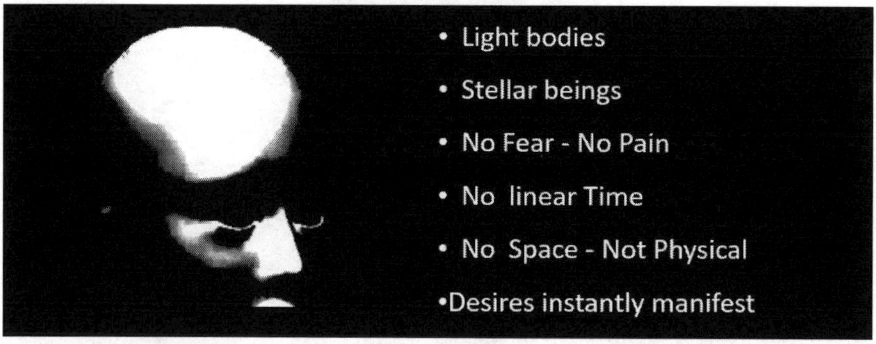

We find even higher frequency increases in the 5th Dimension where physical bodies are not needed to contain consciousness. Existence is light based rather than carbon based as in our physical bodies in the lower dimensions.

This is a realm made up of light bodies and stellar beings where desires instantly manifest and nothing is physical. There is no space and time is not linear and existence is eternal.

Lower frequency emotions like fear, guilt, hatred, and judgment do not exist in this sphere. There is no pain, no sickness, no sorrow or disease and the concept of self-worth is not based on any accumulated material wealth or accomplishment. Essentially it is a world of unconditional love, forgiveness, and acceptance and existence is eternal.

The same identity can exist simultaneously on multiple dimensions. An individual can experience the conscious realm around them in physical reality 3rd Dimensional Consciousness, but a barrier is in place that does not allow them to consciously direct connect with their identity in the higher dimensions.

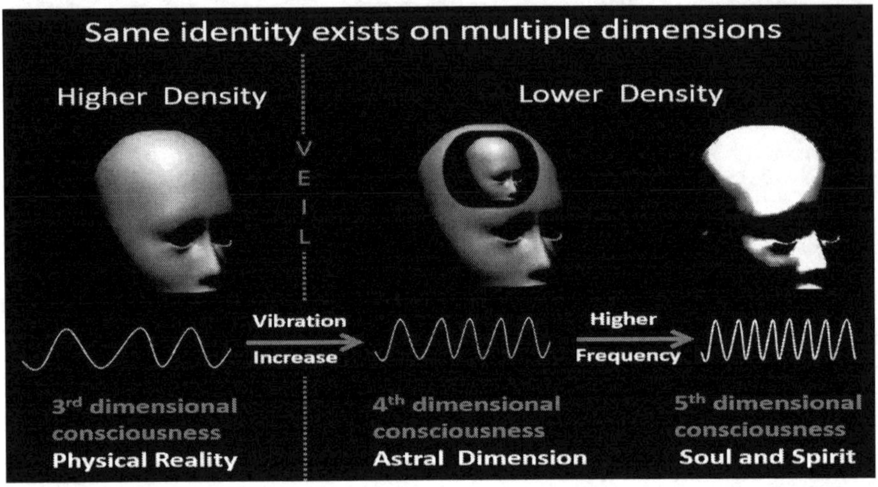

Ancient teaching and religious scriptures talk about a veil that exists between the living and the dead. A chasm that cannot be crossed while in physical reality. As indicated the 5th Dimension is not physical. The frequency in 5th Dimensional Consciousness is so high that a direct connect with a physical human body would burn it up.

Density and Frequency

The concept of density comes into play here. Essentially frequency and density are inversely proportional. As the vibrational frequency increases density decreases.

As we understand our laws of physics, vibration or frequency can be compared to energy. As frequency increases the energy increases and density decreases. For example water has different solidities based on temperature (energy) as illustrated here.

At lower tempatures or lower frequency, water is solid in the form of ice with high density. By increasing the frequency or energy, the ice becomes lower in density and turns to liquid. Continue to increase the frequency with the energy of heat and water turns into a gas which is very lower density.

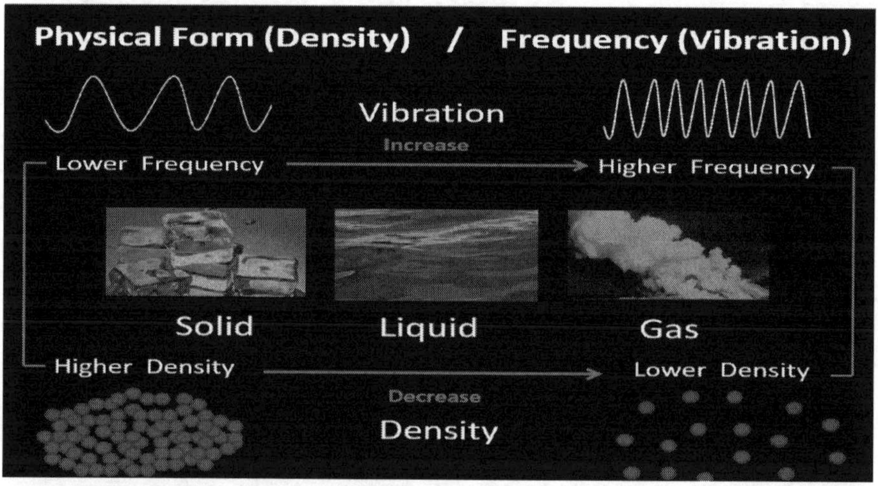

Planetary Magnetic Fields

There's definitely a relationship between the magnetic fields that emanate from a human being and the magnetic fields that surround the earth which effect human consciousness.

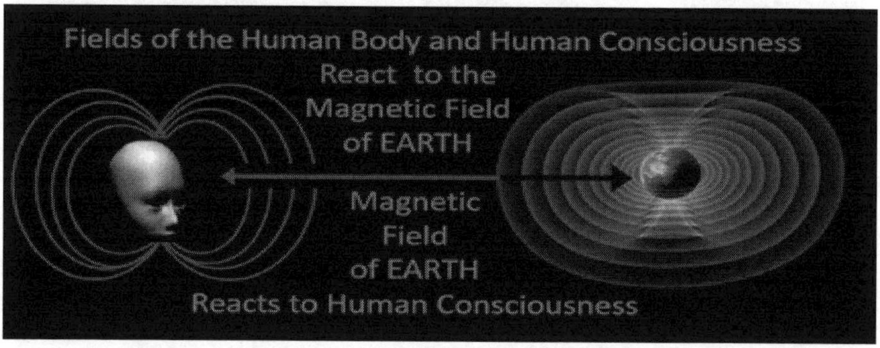

This magnetic field that emanates from the earth fluctuates and is believed to be currently declining to the weakest point that it has been in the last 2000 years. This field effects and controls sleep patterns, dream states, psychic abilities, affects intuition and perception of time which becomes more fluid during periods of weaker magnetic fields.

Schumann Resonance

From the dawn of time, the planet Earth has had what is known as “natural frequency." The Earth's natural frequency is called the Schumann Resonance, which pulsates at a rate of 7.83 hertz. It surrounds and protects all living things on the planet. It was first discovered in 1899 and the ancient teachings refer to it as the heartbeat of the earth. A healthy human body is tuned to this frequency when the Alpha brain waves are in sync with the Schumann Resonance.

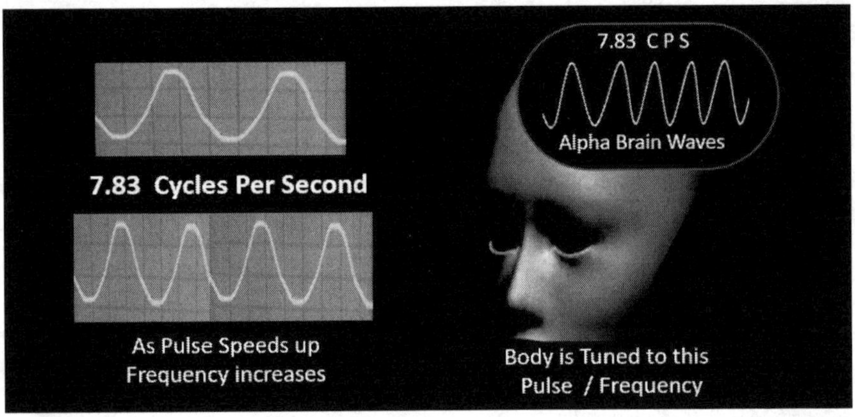

There is some evidence indicating that this residence frequency or pulse may be speeding up which will cause the perception of time to feel like it is moving faster.

Human knowledge

Finally, taking a look at how human knowledge has developed over the last 6000 years gives some insight into exponential explosion of information.

If we could somehow look back to the Sumerian age to about 3800 B.C. and analyze the 6000 year period of time to 1900 we could quantify the accumulated knowledge of human beings over that 6000 year period. Then look at the next 50 years we would find that during the industrial revolution from 1900 to 1950 that accumulated knowledge actually doubled.

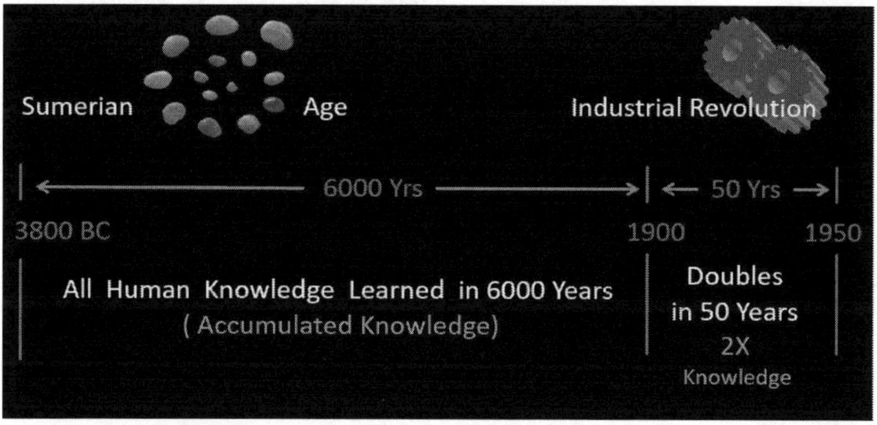

Consider that one current issue of the Wall Street Journal contains more information than the average person learned during an entire lifetime around the turn of century in 1900.

Is human consciousness rising to a higher level?

With the development of technology, computers and most importantly the Internet, human knowledge is exploding.

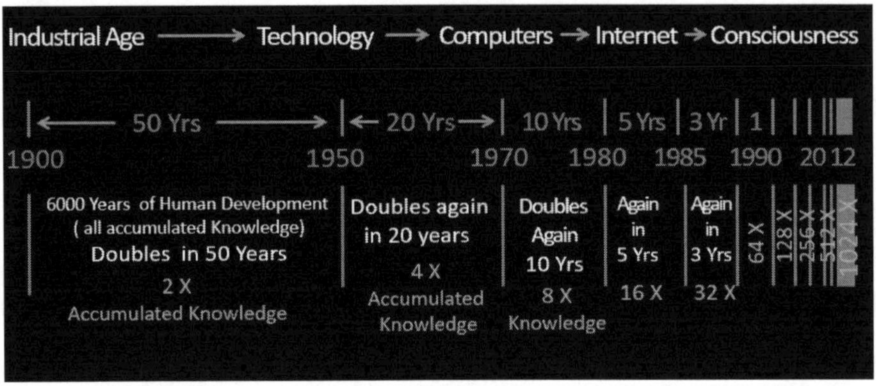

Lets look at the 20 year period from 1950 to 1970, human accumulated knowledge doubled again. Then from 1970 to 1980 which is only a 10 year period and it actually doubled again. From 1980 to 1985 which is only a 5 year period, It doubled again then in the preceding three years it doubled again then in one year it doubled and now is doubling almost on a daily basis, since 1950 it has doubled more than 1024 times. This illustration is hypothetical, but I am sure you get the point.

Take into consideration the massive exchange of information at the speed of light and also the breakdown of cultural and language barriers. You can send an email in English to Europe and someone in France can read it in French. The internet is essentially an electronic form of human consciousness.

END CHAPTER

Chapter 11

6-Layer Model for Anomalous Phenomenon

Shortly after I was hired to be the Manager and Coordinator of the SIP project, I came in possession of a whitepaper known as: " **Incommensurability, Orthodoxy and the Physics of High Strangeness: A 6-layer Model for Anomalous Phenomena" (Vallee and Davis).**

As indicated in Chapter 3, John Schussler had sent it to me, and I studied it intensely. I realized that it was right on the mark as far as investigation and research about UAP phenomenon.

This 6-layer Model for Anomalous Phenomena had become fundamental in my thinking, not only from a standpoint as to how to evaluate case report information (data), but how such case reports should be written, and how the respective data bases should be maintained.

Most investigators write reports from the perspective of their perception of reality as opposed to reporting the data accurately as presented to them, (which we are going to explore in this chapter).

You can read a copy of the paper via a link that is included with the Academic Reference for the paper in the reference section at the end of the book [23].

Please note that the remainder of this chapter is <u>my interpretation</u> of the 6-layer Model for Anomalous

Phenomena, based on observations during my year as manager of the SIP Project and my experiences while investigating cases over the last 30 years.

In 2011, I wrote to Dr. Eric Davis and Dr. Jacques Vallee the authors of the paper and requested permission to reference it in my publications.

You can see below and on the next page their email response to my request:

From: Eric W. Davis [redacted] Sent: Mon 3/21/2011 12:56 PM
To: Richard Lang
Cc:
Subject: RE: Former BAASS/MUFON SIP Project Manager requests permission to reference your white Paper.

Dear Mr. Lang:

Thanks for your very kind message and your recent background info which I am already very well aware of. Dr. Vallee is the lead author of the paper that you are referring to, and I believe that he already sent you a response on my behalf. I see a message from him in my email in-box that I haven't opened yet so that should be it.

Anyway, our paper was published in three different books because it was very popular. The first one is an international forum proceedings volume (published by a university) which is where the paper first appeared because it was an invited paper at an international forum on science, religion and consciousness that was organized by the University Fernando Pesoa in Porto, Portugal.

When citing our paper in any of your publications or online posts, please cite this academic publication as the proper paper reference:

Vallée, J. F. and Davis, E. W., "Incommensurability, Orthodoxy and the Physics of High Strangeness: A 6-Layer Model for Anomalous Phenomena," in *Proceedings of the 2nd International Forum on Science, Religion and Consciousness Oct. 2003*, eds. J. Fernandes and N. L. Santos, Center for Interdisciplinary Study of Consciousness, University Fernando Pessoa, Porto, Portugal, 2005, pp. 225-239.

The forum was held in Oct. 2003 but the proceedings didn't get published until 2005.

Cheers,

Eric W. Davis, Ph.D., FBIS, AFAIAA
Senior Research Physicist
Institute for Advanced Studies at Austin

A link to the paper can be found in the references section at the end of the book [23]

From: [redacted] Sent: Mon 3/21/2011 3:17 AM
To: rlang[redacted]
Cc: ewdavis@[redacted]
Subject: Re: Former BAASS/MUFON SIP Project Manager requests permission to reference y...

Dear Mr. Lang,

Thank you for your kind message and congratulations for your dedicated work. Our paper is in the public domain and free of charge. I do appreciate your requesting permission to quote it, but there is no limitation to its use as long as credit is accurately given to Dr. Davis, myself and the original publication.
For convenience, note that the full text is on my website (www.jacquesvallee.com) both in English and French. Follow the link to "personal research"

I look forward to your future publications on the subject.
Most cordially,
Jacques

In a message dated 19/03/2011 20:33:59 Paris, Madrid, rlang[redacted]writes:

> Mr. Vallee
>
> During the period from February 2009 until January 2010, I served as the Manager and Project Coordinator for the BAASS / MUFON SIP Project. Part of my role was to be the single point of contact for the BAASS / MUFON Relationship.

Incommensurability, Orthodoxy and the Physics of High Strangeness: A 6-layer Model for Anomalous Phenomena"

Layer I:	**Physical**
Layer 2:	**Anti-Physical**
Layer 3:	**Psychology**
Layer 4:	**Physiological**
Layer 5:	**Psychic**
Layer 6:	**Cultural**

Layer I: Physical layer

This first category is the realm of physical reality were most UFO/UAP investigators collect physical evidence, which is documented in accounts that describe an object that:

- Objects that occupy a position in space, consistent with geometry (length x width x height).
- They are aligned with linear time and move as time passes.
- Objects that interact with the environment through thermal effects and produce turbulence moving through physical air space.
- Power output estimates can be derived and calculated.
- When landed, these objects leave indentations and tracks on the ground as well as burn marks (mass and energy figures can be calculated).
- They exhibit light absorption and emission which can be seen, videotaped, and appear in photographic images.
- Sounds emitting from these objects can be heard and recorded (low vibration sounds and whirling sounds).
- Objects can be detected on radar and tracked to determine velocity, trajectory, and flight path.
- They leave material residue consistent with earth chemistry that gives rise to electric, magnetic, and gravitational disturbances which may result in the emission of radiation, electromagnetic fields, and electrogravitic energy.
- May cause effects on the ground producing damage to electric equipment and electronic devices, as well as triggering alarm

systems, turning on automatic lighting systems and activating automated gates and doors.

- Batteries used on electronic equipment and cameras can be completely drained when in close proximity to these objects.
- Objects that have been known to react to high intensity handheld spotlights projected from the ground. There are a number of reports about individuals using high power spotlights to illuminate the bottom of a UAP in close proximity. Hunters that have the million-candle power spot-light in the pickup truck (used to spot dear) have pointed the light beam at a UAP with dangerous consequences.

Layer II: Anti-physical.

Anti-physical for lack of something better to call this, includes objects that may be physical, but they completely defy the current laws of physics in events of very high strangeness.

- The observation of an entity or object that can pass through solid walls or pass through a metal fence. (lower density).
- An object that appears to be plummeting into the solid ground very smoothly without impact or crash, like a rock that is dropped into water (object appears to travel into the side of a mountain without the presence of any opening or doorway).
- Objects that appear to be shrinking in size, growing larger, or changing shape on the spot while being visually observed or becoming fuzzy and transparent while disappearing from view.

- Orbs of light observed dividing into two or more craft, or several of them merging into one object at slow speed.
- Appearing as balls of colored, intensely bright light sometimes pulsating and under intelligent control.
- Objects disappearing at one point and appearing elsewhere instantaneously.
- Triangle shaped crafts converging into one large object.

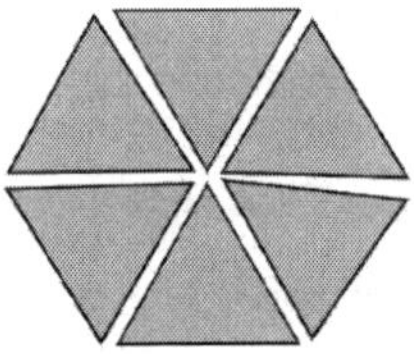

- Objects remaining observable visually, while not detected by radar or not showing in the electronic viewfinder of a video camera.
- Affecting observers who experience missing time (time dilatation). witnesses on their way home from the mall, have a close encounter with a craft at close proximity, that appears to only last for several minutes. Upon arriving at home, they realized that 3 hours have passed since they encountered the object.
- The Silence that accompanies some types of UFO encounters, complete silence which might indicate that time has been stopped, as sound is a frequency based on a unit of time, (for example a guitar string vibrates at a certain frequency and produces the sound).

- Objects in close proximity to automobiles, can cause engines to stop and electrical systems to completely fail. Witnesses indicate that during such an event the engine stops running, all the electric power goes out (headlights, interior dash lights, radio etc.) and cellular phones go dark. After the event suddenly the electric power and lights are back on and the engine is running. Witnesses explain the engine did not start up, it was just running again, (like watching a video, pushing the pause button and then pushing the play button).
- Levitation of the witness or of objects and animals in the vicinity.
- Producing topological inversion or space dilatation (object was estimated to be of small exterior size, but when the observer entered, they experienced a huge interior many times the exterior size.

Investigators do not often report these events accurately because they compare the witness statements to their own perception of reality and ultimately write the report to somehow make sense of the phenomena through their own understanding.

For example: a witness reports that an object approached them as it was coming down through the forest, it went through the trees and then through a metal fence before it landed in the field.

After interviewing the witness, the investigator writes the report indicating that the object approached the witnesses by coming over the trees, flying over the fence and then landing in the field. Thus, a report is filed with very distorted information based on the investigator's lack of understanding about density and multi-dimensional existence.

The desired result would obviously be to report exactly what the witness stated even though it might not make sense to the investigator. Eventually with a more accurate data base this type of phenomena can be investigated and evaluated particularly when you find hundreds of witness's reports that state this exact same thing over time.

Inter-Dimensional Portals

Another element that should be included with the Anti-physical layer, that is not part of the original writing, but remains worth mentioning is Inter-Dimensional Portals.

- Inter-Dimensional Portals opening, (which is often associated with missing persons disappearing in the wilderness).
- Starts with a machine like feeling or noise, sometimes described as a whirling noise, which is the opening of the portal.
- Victims see something blurry (distorted image) in front of them that is interesting and attractive but not scary to them.
- Sometimes these encounters include an entity that is tall, slender and appears to be female, who lures people into the portal.

- Induction through the portal opening is swift and powerful but silent, and sometimes causes shoes and other articles of clothing to be ripped from the individual being pulled in.
- After the event when the portal closes violent weather including thunder and lightning storms occur.

Individuals enjoying the national forests and hiking, sadly do disappear without a trace in alarming numbers. Typically, these individuals are alone when they disappear, having recently walked away from the other members of their group.

Searchers find shoes, boots and articles of clothing near where they were last seen but no trace of a body. Sometimes after several weeks the victim's body is found 10-12 miles away, without rational explanation as to how they got there. Other times they are never seen again,

Inter-Dimensional Portals are typically connected to a geographic area like the one at the Skin Walker Ranch, whereas Abductions are usually associated with some type of craft that is hovering or has landed nearby.

Abduction

Abduction is similar in the sense that individuals are pulled into higher frequency dimensional existence with lower density, but usually confined inside a craft and later returned to their original location. Many experiencers (people that have experienced some kind of abduction phenomena) talk about being in their homes or in close proximity to their house, and some even in their bedrooms, remember being taken away by entities, to be examined on some type of craft and then eventually returned unharmed, with the exception that the psychic affects, may bother them for the rest of their life.

- They hear a low frequency, humming sound.
- They feel a vibration in their body.
- They experience a lack of mobility (paralyzed).
- They are confronted by some type of entities that take them away, sometimes right through the walls or ceilings, into a craft which is usually hovering in the vicinity of where the abduction takes place.
- They indicate the craft is much bigger on the inside than it appeared to be on the outside.
- The buzzing sound is associated with being surrounded by higher frequency energy and actually being pulled into a higher dimension which would explain the reduction of density allowing them to pass through solid objects.
- Experiencers typically will indicate that they were put through some type of a physical examination while on the craft.
- Sometimes they were in the presence of other humans that were taken as well.
- Loss of volition experienced during and after events.

The underlying rationale for being taken varies. Some experiences, particularly females, indicate that they have undergone some type of a gynecological exam and in some cases after the event when returning home, they experience the symptoms of a miscarriage.

Some of the experiencers indicate that they've had multiple abduction encounters, some saying the first time was when they were a little kid walking around in a field near home and then later as a 38-year-old repeating the experience.

Most of the experiencers (abductees) that I've interviewed would tell you that somehow these entities shared with them premonitory information about world events, in a benevolent way. Either by conveying scenes by some form of telepathy or actually showing them images on a screen.

Layer III: Psychology

The third layer has to do with the psychology of the witnesses and the social conditions that surround them. They perceive the objects as non-conventional, but they try to explain them away as common occurrences, until faced with the inescapable conclusion that the object is truly unknown.

Layer IV: Physiological

This has to do with physical effects on the human body and the physiological injuries that occur form close proximity exposure.

- The phenomenon is reported to produce sounds that are heard (beeping, buzzing, humming, sharp/piercing whistling, swooshing/air rushing, loud/deafening roaring, sound of a storm).
- Vibrations sounds felt and heard.
- Burns to the skin often resembling sunburn.
- Partial paralysis typically during encounters.
- Sensation of extreme heat or cold.

- Odors smelled by witnesses, powerful, sweet, or strange fragrance, rotten eggs, sulfurous, pungent, musky.
- Metallic taste in mouth after encounter often described as a copper taste like having pennies in the mouth.
- Prickling sensation to skin.
- Temporary blindness when exposed to the objects intense light.
- Nausea, vomiting dizziness and disorientation.
- Bloody nose or ears; severe headache.
- Difficulty in breathing.
- Loss of volition (not being able to make a clear choice or decision).
- Drowsiness in the days following a close encounter.

Layer V: Psychic

The fifth category of effects can only be labeled psychic because it involves a class of phenomena commonly found in the literature of parapsychology, sighting reports contain accounts of these effects.

- Impressions of communication without a direct sensory channel sometimes known as mental telepathy.
- Poltergeist phenomena: motions and sounds without a specific cause, outside the observed presence of a UAP.
- Maneuvers of a UAP appearing to anticipate the witness' thoughts. This is commonly reported where essentially a witness will indicate that they've gone

out to smoke or let the dog out late at night and they notice a light at great distance and high altitude. After observing the light for a period of time the light seems to anticipate their attention and zooms in on them ultimately being very close to directly over them.

- Premonitory dreams or visions (a dream that appears to give advance notice or warning of some future event).
- Personality changes promoting unusual abilities in the witnesses like clairvoyant abilities.
- Healing - witnesses report recovering from illnesses after encounters and also some indicating that they can heal other people.

Looking at a report where a witness claims that the object could anticipate their thoughts, on the surface this sounds unbelievable, however when these same phrases are articulated in hundreds of sighting reports, it makes an interesting case that such phenomena may have actually occurred:

Layer VI: Cultural

The sixth category could be called cultural. It is concerned with society's reactions to the reports, the way in which secondary effects become generated, and the attitude of members of a given culture towards the concepts.

Secondary effects

- Hoaxes
- Science-Fiction
- Scientific Theories

- Cover-Up
- Media Censorship
- Adverse Publicity
- Sensationalism

Multi-Dimensional Existences - Layers of the Model

Next let us take a look at how the different multi-dimensional existences might relate to the various layers of the Model for Anomalous Phenomena. Hopefully, these illustrations may help to clarify some of the high strangeness associated with UFO/UAP encounters and events, by understanding how density and time is affected in higher frequency Multi-Dimensional existences.

Layer 1: Physical

Starting with Layer 1: Physical as you can see illustrated in the chart below, involves:

- 1st Dimensional Consciousness on the Molecular Level,
- 2nd Dimensional Consciousness on the Biological level and obviously
- 3rd Dimensional Consciousness in Physical Reality.

As discussed, this is the realm where most investigators, collect evidence, analyze data, and write reports based on witnesses' interviews, which pretty much confirm to the laws of physics as most people understand them.

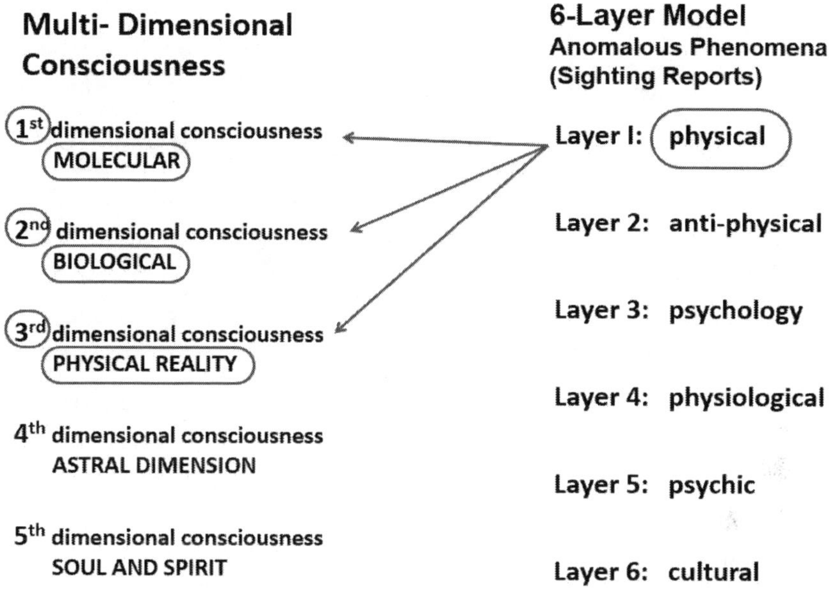

Layer 2: Anti-Physical

As you can see illustrated on the next page, this layer involves Physical Reality 3rd Dimensional Consciousness because objects are described as physical and material, however, they do not confirm to the laws of physics.

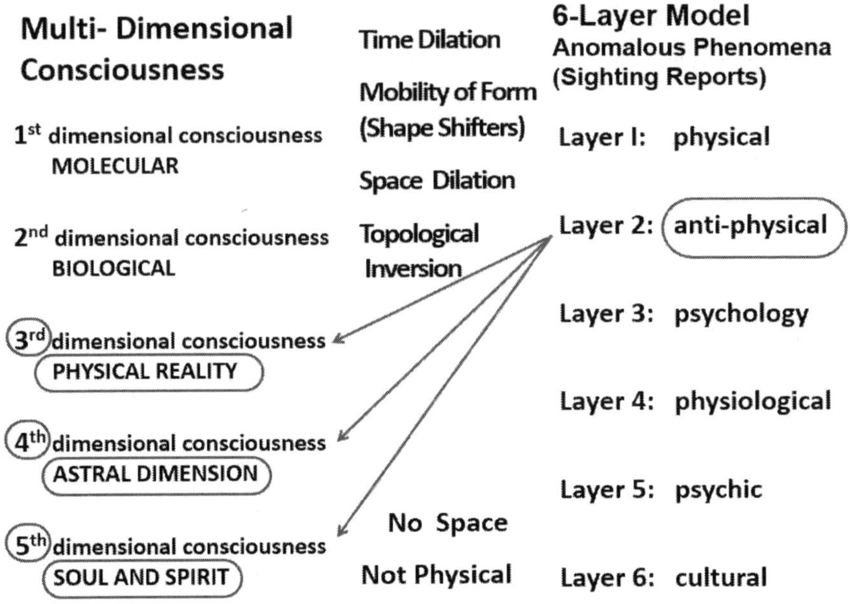

One way to help explain the strangeness might be looking at the higher frequency 4th and 5th dimensional existence and the lower density that is associated with it.

Understanding higher frequency dimensional existence and the fact that time is not linear in these higher frequencies, also helps us to understand how objects can materialize and dematerialize as if they are coming from other times and stopping in this time space reality.

The 5th dimension is not physical, as frequency increases, density decreases, so could it be that during an encounter in the anti-physical layer, that somehow the frequency is increased surrounding the observer and objects pulling them into a higher frequency existence with much low density?

This concept could go a long way to explain how an object could pass through a solid wall. As crazy as it sounds there are literally hundreds of acounts reported where abductees claimed to be taken from their bed straight up through a wall into some kind of craft.

There are many UFO reports claiming that witnesses observed objects have passed through trees, solid walls, fences and even moved into the ground through the side of a mountain.

Layer 4: Physiological
As you can see in the chart on the next page, this layer involves

- 2nd Dimensional Consciousness on the Biological level
- 3rd Dimensional Consciousness in Physical Reality.

This is pretty straight forward where most investigators, collect evidence and analyze data about physiological injuries and other effects on the human body.

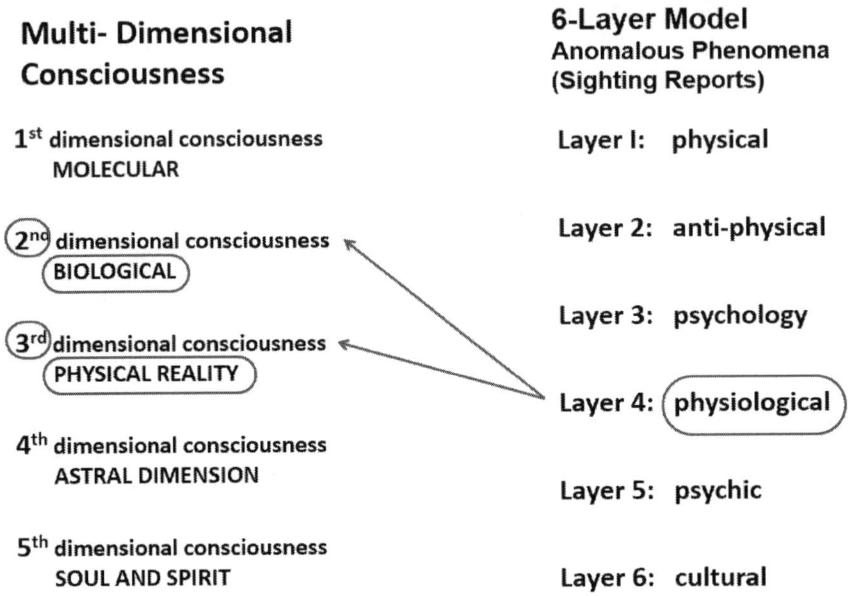

Layer 5: Psychic

You can see illustrated on the next page, the psychic layer involves 4th Dimensional Consciousness (Astral Dimension) and 5th Dimensional Consciousness. (Soul and Spirit)

This is a realm of telepathic communication, premonitary dreams and time dilation. The constraints of physical space and linear time do not affect this layer.

Many of the aspects involving close encounters with entities involved elements of this layer during events and after the events.

Multi- Dimensional Consciousness		6-Layer Model Anomalous Phenomena (Sighting Reports)
1st dimensional consciousness MOLECULAR	Telepathic Communication	Layer I: physical
2nd dimensional consciousness BIOLOGICAL	UAP Reacted to Witness Thoughts	Layer 2: anti-physical
3rd dimensional consciousness PHYSICAL REALITY	Premonitory Dreams	Layer 3: psychology
		Layer 4: physiological
4th dimensional consciousness ASTRAL DIMENSION	No Space	Layer 5: psychic
5th dimensional consciousness SOUL AND SPIRIT	Not Physical	Layer 6: cultural

In the next chapter we will look at report writing, in some part modeled after the six layer model for anomalous phenomena keeping in mind the relationship that the layers have to the dimensional existences.

The ultimate goal is to be writing reports that accurately reflect what witnesses experienced in the reports and that a more reliable data base can be developed to understand this phenomena.

END CHAPTER

Chapter 12

Report writing

Probably the most important component of an investigation is writing the report. Accuracy, attention to detail and unbiased documentation are essential to writing a high quality objective report.

Note that having worked in the corporate world, dealing with individuals on the executive level, I came to understand that the only way they know anything happens is by what is written in the reports. There's an old saying in the corporate world "if it's not in a report, then it didn't happen" You never know where your reports are going to end up and ultimately who might be reading them.

I am fully aware that investigation reports for UFO organizations like MUFON amount to going on a web-based site like CMS and updating the sighting report, by filling in the banks with the investigators comments, findings, photos, and documents.

However if you ever have the occasion to investigate an event for a third party, or help with a law enforcement case, then a more formal written structured format may be necessary. This may also apply when working with an investigative journalist for a news publication or a producer for a documentary TV show. The report format outlined on the following pages in this chapter will serve as a guide for a high quality report that will satisfy most corporate or government agencys.

General Advice to Investigators

During the course of the SIP Project I deployed investigators to work on a wide variety of cases dealing with individuals and witnesses who had submitted sighting reports.

Some of the investigators who were not used to being paid for their work, were often concerned about the outcome of the case, especially when witnesses were submitting false information and were not credable. They were worried about how that would make them look when the report was filed. This is a message I sent to the Star Team Investigators when we first started the SIP Project:

Let me share a couple of things with you: (my two cents worth) You are doing a very fine job and should not ever feel bad or embarrassed!

1) *First of all, we made the decision to deploy on this case and we are solely responsible for any use of funds, resources and expenditure of manpower.*

2) *As an investigator you are not responsible for the outcome of the case.*

3) *You job is to conduct a comprehensive objective fact finding investigation.*

4) *To find the "truth" as to what actually did or did not happen and then report it accurately.*

5) *These cases all have their ups and downs. Just ride it out and see where it goes, take good notes, observe*

everything and write a very accurate objective report when you are done.

6) *Take nothing for granted, never assume anything and check out everything, so someone cannot come behind you and find something you missed.*

7) *It does not matter what direction the case goes as long as you report it accurately.*

8) *Do not endorse the witness (never say whether you believe them or give your opinion in the context of a report)*

9) *Just objectively state the facts!!!*

10) *Use words and phrases like:*
   - *"According to witness testimony"*
   - *"Analysis of the preliminary evidence indicates"*
   - *"We found this to be true"*
   - *"We were not able to substantiate that"*

11) *Do not feel bad if the witness turns out to be a fraud or a hoaxer. If you catch them up in their dishonesty it makes you look good as an investigator.*

12) *Finally, Sometimes these cases will take an unexpected turn and if something comes up later to legitimately discredit a witness you will not look like a fool as long as you made statements of facts.*

Medical Information - Federal HIPAA Laws

Information obtained related to any medical or psychological condition regarding a witness should not be contained in a report as it would be a violation of Federal HIPAA (Health Insurance and Portability and Accountability Act) Laws.

This could open up the door to lawsuits and possible prosecution for the investigator as well as the organization. Many times witnesses and experiencers will share with you details of the medical history or treatment, as well as psychological profile information. Be very careful how you use that information and do not document it in any report format.

Report Documents

Investigation reports should initally be written as a Microsoft Word document. Photos can be included in the document for reference and adding clarity. High resolution copies of the photos should be uploaded to cloud storage and retrieved by internet links included in the written report.

A template word document can be stored on the hard-drive of the investigators laptop as a guide for creating consistent reports.

Once the report is edited and in final form, it should be converted to a **pdf.file** format. The pdf.file format is like taking a photo of a document which can not be changed like a word document. Printed copies can be made from the pdf.file. and the pdf.file can be easily emailed to recipients.

Report Format

(Recommended for UFO/UAP reports)

Organization Logo: (if applicalble)

DATE of Report: (date the report is filed)

CASE # (if applicalble) may also be filed by **Witness Name**

INVESTIGATOR:

- Name (and member number)
- Include City and State of investigator's residence
- Include investigator's cellular phone #

LOCATION of Event: City, State, County and Country

DATE OF EVENT: April 30, 2021

TIME OF EVENT: 2:00 pm EST (indicate time zone)
( ) Exact ( ) Approximate

GPS COORDINATES : Latitude: 39*8'24.21 N
Longitude: 76*32'35.62 W

Automotive and Handheld units can record GPS coordinates which are expressed as the combination of latitude and longitude. (Reference page 147 chapter 8 Equipment)

COORDINATES ACCURACY: (Example) coordinates indicate the location of the witness vehicle on highway where sighting occurred. Object was seen 500 yards to the South East of this location, at an estimated altitude of 1,000 feet.

CASE TYPE: (CAT 1-2-3)

Determine the Vallee classification for the case by using the following questions:

1. Did you witness a strange or unidentified light(s) or object(s) in the sky?
2. Did the light or object land or affect in some way the immediate physical environment?
3. Did any of the witnesses' experience one or more of the following during or after the event?
   I) Observations of beings or entities
   J) interaction with nonhuman entities
   K) episodes of missing time
   L) wounds or trauma to body of unknown origin
   M) taken onboard a craft
   N) telepathic communication
   O) out of body experience
   P) encounter with an unknown animal

CAT 1 = Question 1-Yes Question 2-No Question 3- No
CAT 2 = Question 1-Yes Question 2- Yes Question 3- No
CAT 3 = Question 3 -Yes

Then indicate the clasification and list the specific item(s) in the lists that apply

EXAMPLE

CASE Type: Cat 3

- **episodes of missing time**
- **taken onboard a craft**
- **telepathic communication**

Backround: (include a short summary of report)
Example: Witness was driving home from work a little after midnight on a country road in XYZ county. She noticed a bright light in the sky at high altitude and great distance. After several miles traveling around on the country roads, the light appeared to be approaching and getting close. The witness pulled off the road to observe the light and the light converged directly over her vehicle for a short time and spead off straight up into the sky.

Witnesses Statement: Include any statements the witness made in the original filing of the report first and then add any statements exactly as they were made to the investigators during the investigation.

If video was used during the interview of the witness, include a link to the video for viewing on the internet (cloud storage viewable by other members of your team or organization). If possible a still image can be taken from the video to show the identity of a witness which could be included here.

Investigator Narrative: (write your report here)
This should be a written summary of your investigation, described in your own words in general detail.

Describe your interview with the witnesses and Include your observations. Be specific and include dates and times. Also include the analysis of any data or information that you are able to accumulate during the investigation.

Next address the following 6 catragories

In this next section you can describe the specific Physical evidence you collected and details about the Anti-physical effects associated with your investigation. Be sure to include photos of any measurements or instruments readings you used while obtaining evidence. Then address any insights in the Psychology section and list all physical effects on the body discovered in Physiological section. Finally describe psychic phenomena in the Psychic section. (examples as to type of data and information to be included in each section are in bullet points)

Physical Evidence

This is where you list and describe the physical evidence collected and observed during the investigation and witness interviews.

- Include a physical description of object or craft.
- Include any photos submitted by witnesses.
- Include witnesses estimates as to speed and altitude.
- Descriptions of erratic maneuvering.
- List of any physical evidence collected, like tracks and indentations on the ground as well as burn marks.
- Descriptions of sounds emitting from these objects (beeping, buzzing, humming, sharp piercing, whistling, swooshing air rushing, loud deafening roaring, sound of a storm).
- Effects from a craft on the ground, producing damage to electric equipment and electronic devices (dead batteries, cellular phone damage).
- Interruptions to power or Radio and TV broadcast programing.
- Test results from instruments, if possible, include a photo of the instrument and reading on location.

Anti-Physical Effects

It is important to list Anti-physical effects (there may not be any) but if so include them exactly as they are described by the witnesses. This section includes objects that completely defy the current laws of physics, in events of very high strangeness.

- Object that passes through solid objects.
- Object that passes into solid ground very smoothly without impact.
- Objects that appear to be shrinking in size, growing larger, or changing shape.
- Observers who experience missing time (time dilatation).
- Complete Silence that accompanies UFO/ UAP encounters.
- Objects disappearing and re-appearing instantaneously.
- Levitation of the witness or objects.

Psychology

List any insight as to how the witnesses perceived the event.

- What did they believe actually happened?

Physiological

List any physical effects on the human body including physiological injuries. Photos of burns cuts and bruises should be included here, but not medical information controlled by Federal HIPAA Laws

- Burns to the skin often resembling sunburn
- Partial paralysis typically during encounters
- Sensation of extreme heat or cold
- Nausea,
- Vomiting

- Dizziness
- Disorientation.
- Bloody nose or ears
- Severe headache
- Difficulty in breathing
- Loss of volition

Psychic

List and describe psychic effects (include phenomena commonly known as parapsychology)

- Mental telepathy.
- Poltergeist phenomena.
- Witness claims that UAP appeared to anticipate their thoughts.
- UAP appeared to react to their thoughts.
- Witness describes premonitory dreams or visions.
- Witness claims to develop unusual abilities like clairvoyant capabilities.
- Witnesses report recovering from illnesses after encounters
- Witnesses indicating that they can heal other people.

Cultural

List here any indications of society's reactions to the event in this report. (include Internet links to material)

- Include any Newspaper Articles
- Blog posts
- TV News coverage
- Community outreach
- Statements from Law Enforcement

END CHAPTER

Chapter 13

SOP for Investigative Methodology

Investigating a New Case

Depending on how you learned about the case, will determine who you might have to report to, what kind of report you will be required to file and how you might come to be reimbursed for your expenses. Did a witness contact you? Did an investigative reporter or newspaper ask you to help with a story? Did Law Enforcement or a Search and Rescue organization enlist your services?

Slow Down and Plan Carefully

When you first get a case to investigate, do not grab your gear and jump in your car or head for the airport. Slow down and plan your investigation. Just like when pilots use a checklist to set the controls of the aircraft before starting the engines and taking off, a good investigator uses an SOP (Standard Operating Procedure) as a guide or checklist to plan the investigation.

Study Case Information First

Typically, when an investigator first gets a new case, or they agree to take on a case, they get what is called an **"Intel Kit"** (information package) which is usually emailed and includes everything about the case.

- Who are the witnesses and persons of interest?
- Including photos (for identification)
- Purpose of the investigation.
- Location of events.

- Contact information for witnesses.
- Who to coordinate with for travel and expenses?
- Who to report to and follow up with?

When you get the Intel Kit, the first step is to carefully study and review the case information. Spend a few minutes on the computer and search for any other cases or information that might be related to your case. Make sure to copy the witness contact information so you have it with you.

Contact Witness - Phone Interview

Contact the witness and Interview them on the phone first to verify their contact information and try to validate some details of their report. Most importantly, evaluate their sincerity, are they rational and believable? If they start talking about using your investigation, to help them get on a TV Show, you will know you are wasting your time. Make sure they understand you are the investigator who will be meeting with them.

Schedule Interview - Make Appointment to meet Witness

Make an appointment to meet with the witness and agree on the specific time and exact location where you are going to meet them. Give them your cell number in case there is a mix up about the location.

Google Maps - Check Location of Meeting

Check out the address for the location of your meeting on Google Maps. Click on the satellite view and look to familiarize yourself with the area. If you are meeting at the location of the actual event or sighting, then print a copy of the map and surrounding area for reference during the interview and save a digital copy for your report.

Travel Plan - Coordinate with your Team

Coordinate with whoever you are working with and discuss where you will travel and how long you think you need to stay there. Make sure someone on your team always knows where you are. GPS Satellite Communicators provide Global Connectivity via two-way messaging by connecting to a satellite network as described in Chapter 8 on page 150. It is advisable to have a GPS Satellite Communicator and include the email address of your team members on the device. Upon arriving at the scheduled meeting send a message with coordinates to your team so they know exactly where you are.

Equipment

Pack your equipment based on what you think you will need after talking to the witness and reviewing the case information. In Chapter 8 we illustrated a hard sided waterproof case, that contains basic common-sense equipment that investigators need to carry in the field. This should be the minimum equipment you take with you on any investigation. A laptop like the one shown in Chapter 8 (page 144) is also essential equipment and should always be carried on all investigations.

If you are going to be off road at a location in the forest that is away from your vehicle, then make sure you have basic hiking gear. At minimum, a small backpack to include a

Gore-Tex rain shell, water bottle, good hiking boots, flashlight, headlamp, multi-tool, knife and GPS.

Video Recording an Interview

If you plan to video tape an interview with a witness, which is always a best practice, in Chapter 7 starting on page 123 is a great outline for this procedure. Make sure you have the camera equipment you will need and take some blank evidence forms that are described on page 126.

Personal Safety - Your personal safety is primary over any objective of an investigation. Whenever you go to meet with someone always use caution about the environment you are working in. Do not enter any area that Radiation, Chemical, Biological or Electromagnetic Hazard may be present.

Safety in Numbers - Anytime you can have someone else with you, like another investigator, team member or friend is better than being alone, not to mention that the other person may see something that you missed or provide some unique insight that adds quality to the investigation report. Back in the early days some of the older investigators always said it was good to have some snacks and beverages (water) with you. They advised to never to eat or drink anything that a witness may offer you. Sounds a little paranoid but for what it is worth, I have always followed that advise.

Over the years I have interviewed many people who for the most part were sincere decent people, who were looking for answers. There are a few scammers out there and probably even a smaller number of dangerous individuals, so always be careful and stay in touch with your team members.

END CHAPTER

Chapter 14

Media and the Press

In 1983 the news media in the United States was operated by over 50 individually owned companies that published newspapers and magazines, managed all radio stations and TV broadcasters, they made all the movies, TV shows and gave us the Newspapers at the news stand every day.

Back then I think most of us were comfortable, and we really trusted them. For the most part the people in the entertainment business, TV Newsrooms and the publishers of the newspapers were basically decent honest people.

They were obviously independent of each other and competed for audience ratings and advertising revenues. Over the last 20 years these companies have fallen prey to corporate downsizing profiteering and globalization, much the same way as many other industries and institutions have. There was something unique developing in their understanding of the immense power that Television and video can have on human consciousness.

Subconscious Programming

Humans learn primarily through the visual and auditory senses. It is generally understood that most individuals learn about 85% of what they see and only about 15% of what they hear. That is why teachers, managers, salespeople, and other professionals use PowerPoint presentations. If you stand in front of a group and talk to them, only about 15% of what you say is going to stick with them, but if you show a picture or graphic illustration of what you are talking about, it tends to have a much more powerful affect and enhances the learning process. Not to mention that seeing something in writing tends to be perceived as truth. Remember Billy Graham, preaching on TV, he always holds the Bible up and shows you what it says in writing.

The movie theaters and drive-ins back in the seventies learned to use a technique called subliminal advertising. They had huge projectors in the rear of the theaters. As the film flows through the movie projectors, which is basically a series of pictures being flashed on the screen at about 24 frames a second that gives the illusion of movement. Video cameras now do the same thing except the rate can be as much as 60 frames per second.

Theater companies began to edit the film by inserting a single picture of a Coke or a box of popcorn, on just one frame. So as people sitting in the audience were watching the movie, every few minutes the flashed advertisement would come up on the screen, so fast that consciously they would not notice it. However subconsciously the split-second image recorded in their brain. The bottom line is, during intermission people would be lined up like cattle at the refreshment stand, to grab a Coke or box of popcorn that they were programmed to desire.

The movie industry developed this technique, especially in the production of horror movies. The same concept is used where one single frame will be inserted into the frames of the film, with an image of something visually horrible and grotesque. It is not typically seen consciously but the grotesque image records in the subconscious and creates fear and terror.

Television programs are full of ingenious pharmaceutical advertisements as they use a similar technique. On a commercial you see a nice-looking man and woman walking through a park holding hands, while the words on the screen are describing a new medication that cures some horrible disease that one of them has.

In the background you hear a low frequency voice that is talking extremely fast, almost so fast it is hard to understand. ***Side effects include vomiting, fever, headache, incontinence, double vision, dizziness, seizures, blindness, backpain uncontrolled bowel movements, impotence.***

It is amazing as these ads are run on National Television, literally thousands of people will contact their doctor's office the following week, describing the same symptoms and asking for that medication.

Globalization of the Media

In the past 20 years most of the local newspapers, radio stations and television broadcasting companies have been bought out by multinational corporations.

Powerful corporate entertainment and news agencies now control everything we see on television, hear on radio and read in the newspapers. An industry that was once owned by more than 50 independent companies is now controlled by only 6 corporations.

The following list shows the 6 powerful multinational media companies. Under each corporate name, the assets the company controls are listed in billions and the media businesses that they own, and control are listed by bullet point.

Comcast

(74.51 Billimon)

- NBC
- MSNBC
- Universal Pictures
- Focus Features
- USA Network
- DreamWorks Animation
- Bravo
- CNBC
- The Weather Channel
- Syfy
- NBCSN
- Golf Channel
- Esquire Network
- 26 television stations in the United States cable networks

DISNEY

(52.46 Billion)

- ABC
- ESPN
- PIXAR
- MIRAMAX
- MARVEL STUDIOS
- Touchstone
- Marvel Entertainment
- Lucasfilm
- Walt Disney Pictures
- A&E Lifetime
- 30 radio stations, music, and book publishing companies
- Longstanding partnership with HEARST CORPORATION, which owns additional TV stations, newspapers, magazines

Time Warner

(28.11 Billion)

- CNN
- HBO
- Cinemax
- HLN
- NBA TV
- TBS
- TNT
- Turner Classic Movies
- Warner Bros. Pictures
- Castle Rock
- Warner Bros. Interactive Entertainment
- New Line Cinema

21st Century Fox

(28.98 Billion)

- Fox Broadcasting Company
- Fox News Channel
- Fox Business Network
- Fox Sports 1, Fox Sports 2
- National Geographic
- Nat Geo Wild,
- FX Movie Channel
- Fox Sports Networks
- 20th Century Fox
- Fox Searchlight Pictures
- Blue Sky Studios.

NEWS-CORP

- Wall Street Journal
- New York Post
- Barron's Magazine
- SmartMoney Magazine
- Book publisher HarperCollins

CBS Corporation

(13.88 billion)

- CBS Television Networks 30 television stations
- Showtime
- Smithsonian Channel
- NFL.Com
- 60 Minutes
- CBS Sports Network
- CBS Radio which has 130 stations
- Publisher Simon & Schuster

VIACOM

(13.27 billion)

- Paramount Pictures
- Nickelodeon
- Nick at Nite
- TV Land
- VH1
- BET
- CMT
- Comedy Central
- CMT

These 6 corporations and the individuals that operate them, control everything we happen to understand about the world we live in. They make all the movies and produce all the TV series. They have complete control of all content in the public view, as well as the powerful subconscious tactics they use to control the public perception.

One thing for sure is they absolutely do not want you to know anything about UFOs, Extraterrestrial Space Travel or Aliens!

The PRESS and UFO Investigators

The first MUFON symposium I attended was in 1999 at a hotel in Washington DC. I remember being so excited about the many interesting people there from all over the world. Writers, speakers, documentary producers, researchers, scientists, and world class investigators were in attendance. I would guess that there were at least 100 people there educated to the level of PHD. I attended the state directors meeting and received an award for an investigation I had conducted that year. (pictured in Chapter 3)

I was living in Virginia at the time. I drove with one of the other investigators who was the State Director of Virginia at that time to Washington DC and stayed at the hotel where the symposium was. I remember when we pulled up in the drive-through at the hotel, there was some guy out there walking around with a big tinfoil hat on his head acting kind of goofy. He was obviously positioned there to embarrass and discourage the individuals who were attending the UFO symposium.

My point in documenting this is that after the symposium was over, we returned home and noticed that one of the Washington newspapers published an article about the UFO Symposium. Interestingly of all the professionals and PhDs that attended the symposium, you can guess whose picture was on the front page. Yes, the goofy guy with the tinfoil hat was pictured on the front page of the newspaper and as I remember the article was not very complimentary, talking about the conspiracy theory people getting together at the conference center.

I learned early on that the media and press are not your friends when conducting UFO research. Having said that, knowing how to deal with them and not letting them take advantage of you is very important.

TV Shows and Documentaries

Generally speaking, the rules for producing television shows and documentaries in the United States about UFOs require that the producers interview at least one participant that blatantly debunks what has been presented. I am not sure there is any research that indicates any specific State or Federal laws pertaining to this, but generally the networks and the media companies in the USA all follow the same rules.

For example, if they produce a one-hour documentary about a UFO case, first the investigators present information, pictures and testimony about an event or case they investigated. Then at least 25% of the remaining time will be allocated to someone with an adversary view, which usually comes at the end.

So, during the first part of a typical UFO television show the investigators and witnesses enthusiastically present the case and evidence, then during the last 15 minutes, a guy wearing a bowtie comes on claiming to have scientific credentials, and basically discredits the witnesses and debunks everything the investigators said.

Often investigators participating in network television shows are separated and isolated in a greenroom while being interviewed. They can communicate with the commentator through an earpiece so that they can hear the commentator's questions but cannot see or hear what the other people on the show are talking about. Figuratively speaking it is pretty much like being blindfolded and then pushed off a cliff.

My advice for doing TV shows, unless you have a lot of experience dealing with media production companies, **do not do it!** But if you do, make sure you know who you are dealing with, what the rules are, and what kind of environment that you will be filming in.

MEDIA outside of the USA
In all fairness I would like to say the rules about filming in Canada are much different. In 2012 and 2013, we flew to Toronto a number of times to work with **Discovery Channel Canada** on a documentary series about UFOs that was called "**Close Encounters**".

The producers created two seasons of “Close Encounters”. I was filmed and interviewed in several episodes of the first Season and the following year invited back to work on Season 2 (pictured here). We worked out of a studio in Toronto, and it was really fun. The shows ultimately aired in Canada, Europe, and the United States

As indicated, I have worked extensively with Discovery Channel filming here in the United States. I can honestly say that working with Discovery Channel Canada was a much better experience. The producers in Canada were considerably more open minded and objective about the subject material.

The rules in Canada are quite different, they apparently have no obligation to utilize the adversary witnesses or debunkers in any part of the shows and documentaries.

In Mexico the subject of UFOs is much more mainstream and accepted by the media (and the general public). Mexican television newscasters seem to be much more open minded and objective about the subject and have broadcasted some incredible video and pictures to the public.

What to watch out for and avoid

There was an old saying about the press and media that goes something like this: "good publicity is good and even bad publicity can be good if they spell your name right". I am not so sure how relevant that is in this digital age, but saying the wrong thing to a newscaster can be really devastating.

If you get an invitation to be interviewed on radio or a podcast some things that I have learned that you might want to watch out for and avoid if you can.

When they interview you or write about you, they will tend to characterize you in a negative way. Words like **Conspiracy Theorists**, UFO **Enthusiasts** or **Believers** are often used to shed a negative light on the work that you might be doing.

The term **Conspiracy Theory**, which is one of the most powerful of ways to discredit someone, was actually a term that was devised by the CIA back in the mid-1960s, after the assassination of JFK. Anyone who dared to question the official government account of the assassination was immediately characterized as a conspiracy theorist, and it really worked well.

Often reporters will refer to you as a UFO Enthusiast, a Believer or UFO Fan. When that happens, you have got to stop them dead in their tracks and correct them.

- **Enthusiasts** are people that go to auto racing events
- B**elievers** are religious people that go to church,
- **Fans** are people that are interested in sports figures and Hollywood celebrities.

An **Investigator** is a person who carries out a formal inquiry or investigation that often may be involved in scientific studies for research.

Remember, as a professional investigator you normally engage in the following activities, which are all you should be talking about:

- Interview Witnesses
- Establish Sequence of Events
- Evaluate Circumstances
- Assimilate Facts
- Examine Data
- Photograph and Measure Physical Evidence
- Write Objective Factual Reports

So, when being interviewed do not ever answer questions about what you think or how you believe, stick to the facts.

Widespread interest in UFOs

It is interesting to note how the public in general feels about UFOs and extraterrestrials. In 2009 AOL conducted a poll that involved about 495,000 people. The participants in the poll were asked two questions:

1) **Do you think that extraterrestrials have visited Earth?**
    - **73% answered YES.**
    - **17% answered NO.**
    - **10% were undecided.**

2) **Do you think the government is trying to cover up the existence of extraterrestrial life?**
    - **76% answered YES.**
    - **15% answered NO.**
    - **9% were undecided.**

The bottom line is that working with the mass media might be an uphill battle and challenging, but the public interest definitely is on your side.

END CHAPTER

Chapter 15

Is Disclosure on the Horizon?

In 1999 a document was published in France entitled *UFOs and Defense: What must we be prepared for?* This ninety-page report is the result of an in-depth study of UFOs, especially addressing questions of national defense. The study was carried out in France over several years at the Institute of Advanced Studies for National Defense (IHEDN) and through COMETA which stands for "Committee for in-depth studies and involved both government and military officers. [24]

The report officially recognizes Colonel Philip Corso and concludes that his testimony (book) is most likely fully accurate and conveys the real situation in the United States Government.

The report explores motivations of extraterrestrial visitors such as protection of planet earth against the dangers of nuclear war, suggested by frequent events of UFOs flying over nuclear missile sites.

The report examines the official behavior of different nations and focuses on the probability that secret, privileged contacts have been established with the United States (by contacts meaning direct relationships with extraterrestrial civilizations). Plus suggesting that since 1947 and the Roswell event, a policy of ever-increasing secrecy seems to have been applied to protect US Military technological research and back engineering projects (using recovered

UFO wreckage) that has been controlled exclusively by the Military Industrial Complex.

As explained in Chapter 5, the concept of Disclosure is not as much about UFOs, Aliens and Extraterrestrials as it is about unwrapping of the secrecy surrounding the Military Industrial Complex, not to mention the advanced technology that has been developed with black project money used at the expense of US taxpayers, ultimately allowing these corporations to make billions of dollars for their own uses.

This is much vaster than most people would expect, even those of us who have spent many years involved in this research. To put this in perspective we acquired the following list of UFO crashes that occurred where wreckage and alien bodies have been recovered.

Date - **Location** - Bodies

July 4, 1947 **- Roswell, New Mexico** - 4 Bodies

Oct 1947 - **Paradise Valley Arizona** - 4 bodies

Feb 1948 - **Aztec, New Mexico** - 12 Bodies

July 1948 - **Mexico, South of Laredo** - 1 Body

1949 - **Roswell, New Mexico** -1 ET Living

1952 - **Spitsbergen, Norway** - 2 Bodies

Aug 1952 - **Ely, Nevada** - 16 Bodies

Sept 1950 - **Albuquerque, New Mexico** - 3 Bodies

April 1953 - **South West, Arizona** - No Bodies

May 1953 - **Kingman, Arizona** - 1 Body

June 1953 **- Laredo, Texas** - 4 Bodies

July 1953 - **Johannesburg, South Africa** - 5 Bodies

Oct 1953 - **Dutton, Montana** - 4 Bodies

May 1955 - **Brighton, England** - 4 Bodies

July 1957 - **Carlsbad, New Mexico** - 4 Bodies

1961 - **Timmensdorfer, Germany** - 12 Bodies

June 1962 - **Holloman, AFB, New Mexico** - 2 Bodies

Nov 1964 - **Ft. Riley, Kansas** - 9 Bodies

Dec 1965 - **Kecksburg Pa** -No Bodies

Oct 1966 - **North West, Arizona** - 1 Body

July 1972 - **Morocco, Sahara Desert** - 3 Bodies

July 1973 - **North West, Arizona** - 5 Bodies

Aug 1974 - **Chihuahua, Mexico** - No Bodies

May 1976 - **Australian Desert** - 4 Bodies

June 1977 - **North West, Arizona** - 5 Bodies

April 1977 - **South West, Ohio** -11 Bodies

Aug 1977 - **Tabasco, Mexico** - 2 Bodies

May 1978 - **Bolivia** - No Bodies

Nov 1988 - **Afghanistan** - 7 Bodies

May 1989 - **South Africa** - 2 ET Living

June 1989 - **South Africa** - 2 ET Living

July 1989 - **Siberia** - 9 ET Living

Sept 1990 - **Megas Platonos, Greece** - No Bodies

Nov 1992 - **Long Island, New York** - No Bodies

If even half of this list is accurate and I am reasonably sure it is, then you can see these guys have been remarkably busy recovering wreckage, doing extensive research, and making billions of dollars while developing technology that is literally out of this world.

So, what is going on?

Over the last 3 years it seems that the Pentagon is starting to release information about the US Government's involvement with UFO/UAP activity.

Remember most of this research, and the technology that has been developed, was facilitated through private multi-national corporations and defense contractors, as opposed to working directly in government agencies and on military bases, which obviously makes their research very private and exempt from FOIA requests.

According to information obtained in late 2020 through a FOIA request, that was submitted to Defense Intelligence Agency, The Pentagon has admitted to testing wreckage recovered from UFO crashes, indicating some of this material was placed with a defense contractor for analysis and storage in "specialized facilities". The documents revealed the tests were carried out by Bigelow Aerospace, in the underground facility at Las Vegas, and that the company performs private contract work for the Department of Defense.

The key words here are: *"private contract work for the Department of Defense".* Just remember back in Chapter 4,

BAASS did not share findings or test results with us as the deal clearly stated that BAASS was a private organization, and their research was proprietary information. I am sure that most of the other defense contractors and technology research corporations operate the same way.

In the next few pages, we can start to look at what the Government is going to disclose about its official involvement in the UFO/UAP activity, which is good.

On the other hand, the Military Industrial Complex, including the defense contractors and research corporations are going to be disinclined (not in any hurry) to release their private proprietary information. Think about all the money they have made using funds that were funneled to them through black budgets and secret government spending programs. The public is going to want accountability for the public funds and taxes, as well as disclosure of advance technology that was paid for with US tax dollars and hidden from them. This will undoubtedly open the door to massive lawsuits and in some cases with legal action against the officers of the corporations.

Declassification and Adverse Consequences

A great example about declassification and the resulting adverse consequences reminds me of when the secret underground bunker underneath the Greenbrier Hotel in West Virginia was declassified. This was a state-of-the-art underground facility built to house 300 members of congress in the event of a nuclear strike on Washington, DC.

The bunker was constructed under top secret orders during the Eisenhower administration (completed construction in 1962) with funding from secret government black project funds. It was done during a building expansion project when they added a new wing on the hotel above ground (bunker was secretly being built underneath the new wing). It operated for about 30 years in a state of stand-by readiness to accommodate the Congress if necessary.

The Bunker was a massive expansion to the hotel building (addition of over 112,544 square feet underground) and a secret staff manned the bunker 24 hours a day. The money to build all this and pay the employees working underground was laundered through the Railroad company that serviced the town (to keep everything secret).

The Washington Post ran an article in 1992 disclosing the existence of the bunker (blowing their cover) and ultimately the facility was declassified.

Since the Bunker and the employees working there did not exist (officially) so tax paying obligations were obviously ignored. However, after the declassification, the issues about taxes that had not been paid (including 30 years' worth of property tax) immediately came up and was investigated by the state and county governments, which ended up with law-suites, tax liens and plenty of dirty laundry to clean up there.

I am sure that the officers of the Military Industrial corporations involved already know this is going to be a nightmare for them when and if they come clean with all they have been involved with. They will likely hang on to the private proprietary information and patent rights they own for as long as they can.

With that all said, the Pentagon has been dripping out bits of interesting information over the last 3 years.

Washington Post - December 2017

The Pentagon's secret UFO program was exposed on December 16, 2017, by the Washington Post in an article titled: *"Head of Pentagon's secret 'UFO' office sought to make evidence public"*, written by Joby Warrick.

This article was about an intelligence officer who arranged to secure the release of unusual videos in the Pentagon's secret vaults: raw footage from encounters between fighter jets and Anomalous Aerial Vehicles (AAV) all taken from fighter jet cockpit cameras, showing pilots struggling to lock their radars on oval-shaped vessels.

According to the article, an internal Pentagon memo requesting that the videos be cleared for public viewing, argued that the images could help educate pilots and improve aviation safety.

New York Times - December 2017

Then an article published in the New York Times by Helen Cooper, Ralph Blumenthal and Leslie Kean, *"Glowing Auras and 'Black Money': The Pentagon's Mysterious U.F.O. Program.* Bottom line is the existence of the program, known as the Advanced Aviation Threat Identification Program (AATIP), was confirmed officially for the first time by the Pentagon in these articles. There was Video embedded (included links) in the articles that show fighter jet cockpit camera recordings.

November 2004 USS Nimitz encounter disclosed

Some of the Videos included with in these articles show encounters between Navy F-18 Super Hornets and AAVs

The USS Nimitz Carrier Strike Group, which included the nuclear-powered aircraft carrier and the missile cruiser USS Princeton, were conducting a series of drills while located in the Pacific Ocean, about 100 miles southwest of San Diego, California.

As indicated, AAV is a term the military prefers to use as opposed to UFO or UAP. The Princeton's radar team had been identifying numerous AAVs appearing at altitudes greater than 80,000 feet, far higher than commercial or military jets typically fly, and they began detecting instances in which the AAVs dropped with astounding speed to lower, busier commercial airspace.

On November 14, 2004, Two F/A-18F Super Hornet fighter jets from the Nimitz, that were already airborne doing training, received an order from an operations officer (air controller) aboard the USS Princeton. They were told to stop their training maneuvers and proceed to new coordinates to investigate an AAV.

The Fighter Pilot of the USS Nimitz Carrier Strike Group and his wing man (who was actually a female fighter pilot) were vectored to the location of possible AAVs. While enroute they were asked if they were carrying live weapons and they replied that they were not.

The Fighter Pilot reported that he saw an unknown object, white and oval like a Tic-Tac, hovering above an ocean disturbance. He described it as about 40 feet long, shaped like a Tic-Tac candy and with no obvious means of propulsion: "It's white. It has no wings. It has no rotors".

As he flew around it, he says the craft ascended and came right at his plane: "All of a sudden it kind of turns and rapidly accelerates, beyond anything I've seen, crosses my nose, and…it's gone. Several videos of these Tic-Tac's filmed form the infrared cockpit cameras, in the fighter jets have been released by the Defense Department's Advanced Aerospace Threat Identification Program. (AATIP).

Note: This information is authentic, but it is only the tip of the iceburg. The Public disclosure from these two articles has helped in bringing the UFO subject back to the front burner in the public view. Links to the articles can be found in the references at the end of the book [25]

Most recently and of extremely high significance is the mandate in the appropriations bill on the next page.

Intelligence Authorization Act for Fiscal Year 2021

Part of the $2.3 trillion dollar appropriations bill signed on December 27, 2020, mandated that a report detailing what the government knows about UFOs be released this year.

The legislation, contained the Intelligence Authorization Act for Fiscal Year 2021, which had a provision "**Advanced Aerial Threats.**" This stipulation mandates that the director of national intelligence work with the secretary of defense on a report detailing everything the government knows about Unidentified Flying Objects also known as "Unidentified Aerial Phenomena" or "Anomalous Aerial Vehicles."

The report must include "detailed analysis of Unidentified Aerial Phenomena data and intelligence" gathered by the Office of Naval Intelligence, the Unidentified Aerial Phenomena Task Force, and the FBI.

When Trump approved the spending package in December, a 180-day countdown began, giving intelligence officials until June 27th to deliver lawmakers their report.

Office of National Intelligence - John Ratcliffe

Former director of the Office of National Intelligence, John Ratcliffe said in a recent interview on FOX news "Frankly, there are a lot more sightings than have been made public," and "There are instances where we don't have good explanations for some of the things that we've seen," "and when that information becomes declassified, I'll be able to talk a little bit more about that."

Ratcliffe indicated “We are talking about objects that have been seen by Navy or Air Force pilots, or have been picked up by satellite imagery, that frankly engage in actions that are difficult to explain, movements that are hard to replicate, that we don’t have the technology for and are traveling at speeds that exceed the sound barrier without a sonic boom.” “and I think the information that is being gathered will be put out in a way the American people can see.”

Vice-chair of the Intelligence Committee - Marco Rubio
Sen. Marco Rubio (R-Fla.), Vice-chair of the Intelligence Committee, said the prospect that something otherworldly is behind the flying objects does not concern him as much as the idea that a U.S. adversary could be making secret technological advances.

According to Rubio, “The bottom line is if there are things flying over your military bases and you don’t know what they are because they aren’t yours and they are exhibiting potentially technologies that you don’t have at your own disposal, that to me is a national security risk and one that we should be looking into,”

Hopefully by June 27 we will see some significant information coming from this report.

Richard Lang - Virginia - USA

References

A copy of this reference section with
live hyperlinks can be found on my blog

from "the Dark Side of the Looking Glass"

https://fromthedarksideofthelookingglass.blogspot.com/

[1] Largest UFO reporting centers in the USA:

THE NATIONAL UFO REPORTING CENTER
http://www.nuforc.org/ and

MUFON (Mutual UFO Network)
http://www.mufon.com/reportufo.html

[2] Embry-Riddle Aeronautical University is a private university in the US specializing in aviation and aerospace engineering. It teaches the science, practice, and business of aviation and aerospace, as well as flight training, ground school, and technical training programs. Located in Daytona Beach Florida For more information about the University information see http://www.erau.edu/

[3] The Mutual UFO Network (MUFON) is an American non-profit organization that investigates cases of reported UFO sightings. It is one of the oldest and largest UFO investigative organizations in the world. For more information see http://www.mufon.com/

[4] Samuel M. Yount, president of Yount Investigations, Inc. A Premier Private

Investigation firm serving clients for over 40 years, located in Charlottesville, Virginia. For more information see http://www.yountinvestigations.com

[5] Filer's Files: MUFON Skywatch Investigations - #18 (May 7, 1999) by George A. Filer - MUFON Eastern Director Virginia Black Helicopter - NO LIGHTS, Richard Lang State Section Director of MUFON Virginia reports the strangest thing occurred at the lake where we live. For more information see http://www.netowne.com/ufos/filers/filers/filers18.htm

[6] Filer's Files: MUFON Skywatch Investigations - #36 (September 9, 1999) by George A. Filer - MUFON Eastern Director, MUFON OF VIRGINIA-Sighting & Alleged Abduction - VINT HILL FARMS BASE -- Richard Lang, Assistant State Director VA MUFON writes, "We recently investigated a sighting and alleged abduction which took place on a now closed military base in Northern Virginia." For more information see http://www.netowne.com/ufos/filers/filers/filers36.htm

[7] Albemarle County Sheriff's Office Reserve Division the Reserve Division provides law enforcement back up and support for all Albemarle County Sheriff's Office functions, including community support activities. Providing many valued services at little to no cost to the taxpayer. For more information see http://www.albemarleso.org/Nreserve.html

[8] Central Shenandoah Criminal Justice Training Academy is a state Law Enforcement Training facility located in Weyers Cave, Virginia, one of ten regional academies within the State of Virginia preparing individuals for careers in local law enforcement. http://www.cscjta.com/

[9] MUFON UFO Journal - Richard Lang the newest member of STAR Team **-** August 2008 Volume No. 484 page 15 see: photo and article

[10] MUFON UFO Journal - Richard Lang appointed as STAR Team Coordinator, who would facilitate communication and deployment - May 2008 Volume No. 485 page 22 see Director's Message - MUFON STAR Team Coordinator

[11] UFOs over Earth - Discovery Channel TV Series - "Fayetteville Incident" produced by Creative Differences initially aired 2008 you can watch the show on YouTube at the following links:

1) Fayetteville Incident - Part 01 of 04
   http://www.youtube.com/watch?v=jQ4h0j0SoHI
2) Fayetteville Incident - Part 02 of 04
   http://www.youtube.com/watch?v=plkK42K-Wac
3) Fayetteville Incident - Part 03 of 04
   http://www.youtube.com/watch?v=oSQMfuvRVBc
4) Fayetteville Incident - Part 04 of 04
   http://www.youtube.com/watch?v=0KfHiBOW0Oo

(Note the author or publisher take no responsibility for the posting of these links, they are provided solely as the result of an internet search for the titles)

[12] UFOs over Earth - Discovery Channel TV Series - "The Bucks County Flap" produced by Creative Differences initially aired 2008 you can watch the show on YouTube at the following links:

1) Bucks County Flap Part 01 of 05 http://www.youtube.com/watch?v=x_Cwjgv-Ii4
2) The Bucks County Flap Part 02 of 05 http://www.youtube.com/watch?v=L_sFWsluvSU
3) The Bucks County Flap Part 03 of 05 http://www.youtube.com/watch?v=lA17Y5ZnN70
4) The Bucks County Flap Part 04 of 05 http://www.youtube.com/watch?v=qQYrrP9bLIg
5) The Bucks County Flap Part 05 of 05 http://www.youtube.com/watch?v=JmTdzx_H2UA

(Note the author or publisher take no responsibility for the posting of these links, they are provided solely as the result of an internet search for the titles)

[13] Bigelow Aerospace https://bigelowaerospace.com/
From Wikipedia, the free encyclopedia, Bigelow Aerospace
https://en.wikipedia.org/wiki/Bigelow_Aerospace

[14] Elaine Douglass blog - May 12, 2011
"The Gagged-for-life Star Team Confidentiality Agreement"https://www.ufodigest.com/article/the-gagged-for-life-star-team-confidentiality-agreement/

[15] CBS "60 MINUTES" TV show aired May 28, 2017.
Correspondent Lara Logan's interview with Robert Bigelow
"Bigelow Aerospace founder says commercial world will lead in space" http://www.cbsnews.com/news/bigelow-aerospace-founder-says-commercial-world-will-lead-in-space/

[16] Keith Basterfield Blog Publisher *"Unidentified Aerial Phenomena - scientific research"* an examination of aspects of Unidentified Aerial Phenomena (UAP) from a scientific perspective.

https://ufos-scientificresearch.blogspot.com/

Keith has been interested in the subject of unidentified aerial phenomena (UAP) since the 1960's. Over the years he has held a variety of roles in numerous Australian state, national, and international organizations, studying the phenomena. His work has appeared in various state, national and international newsletters, magazine and journals about UAPs.

[17] Blue Blurry Lines - Examining the Mystery and Myths of Unidentified Flying Objects

- ***The Pentagon UFO Program's Secret Partner*** By Roger Glassel and Curt Collins, https://www.blueblurrylines.com/2020/03/the-pentagon-ufo-programs-secret-partner.html
- **AAWSAP-BAASS-MUFON Document Collection 2008 - 2009** https://app.box.com/s/x54czwgmwfd5fph50527xw9zvma1yy91

[18] The Day After Roswell

by Philip J. Corso and William J. Birnes

The Day After Roswell is an American book about extraterrestrial spacecraft and the Roswell UFO incident. It was written by United States Army Colonel Philip J. Corso, with help from William J. Birnes, and was published as a tell-all memoir by Pocket Books in 1997, a year before Corso's death. The book claims that an extraterrestrial spacecraft crashed near Roswell, New Mexico, in 1947 and was recovered by the United States government who then sought to cover up all evidence of extraterrestrials. The book appeared on The New York Times Best Sellers List for several weeks,

[19] "**Sent by Eisenhower to Meet an ALIEN**, an Amazing Confession of Ex Military" by Richard Dolan https://www.youtube.com/watch?v=d4O89Xtt47Q

[20] Project Blue Book From Wikipedia, the free encyclopedia, https://en.wikipedia.org/wiki/Project_Blue_Book

[21] NORAD From Wikipedia, the free encyclopedia, https://en.wikipedia.org/wiki/North_American_Aerospace_Defense_Command

[22] 1080p HD From Wikipedia, the free encyclopedia https://en.wikipedia.org/wiki/1080p

[23] Vallée, J. F. and Davis, E. W., "Incommensurability, Orthodoxy and the Physics of High Strangeness: A 6-Layer Model for Anomalous Phenomena," in Proceedings of the 2nd International Forum on Science, Religion and Consciousness Oct. 2003, eds. J. Fernandes and N. L. Santos, Center for Interdisciplinary Study of Consciousness, University Fernando Pessoa, Porto, Portugal, 2005, pp. 225-239. See full text http://www.jacquesvallee.com/ both in English and French. Follow the link to "personal research" https://www.jacquesvallee.net/wp-content/uploads/2018/11/Incommensurability_Orthodoxy_and_the_Phy.pdf

[24] The French Report on UFOs and Defense:
A Summary By Gildas Bourdais cometa.pdf (cufos.org)

July 16, 1999 an important document was published in France entitled, UFOs and Defense: What must we be prepared for? ("Les Ovni Et La Defense: A quoi doit-on se préparer?"). This ninety-page report is the result of an in-depth study of UFOs, covering many aspects of the subject, especially questions of national defense. The study was carried out over several years by an independent group of former "auditors" at the Institute of Advanced Studies for National Defense, or IHEDN, and by qualified experts from various fields. Before its public release, it has been sent to French President Jacques Chirac and to Prime Minister Lionel Jospin

[25] Washington Post / New York times December 2017

- **Glowing Auras and 'Black Money': The Pentagon's Mysterious U.F.O. Program**
  https://www.nytimes.com/2017/12/16/us/politics/pentagon-program-ufo-harry-reid.html
  Note: video is embedded in this article that shows an encounter between a Navy F-18 Super Hornet and an unknown object. It was released by the Defense Department's Advanced Aerospace Threat Identification Program.
- **Head of Pentagon's secret 'UFO' office sought to make evidence public**
  https://www.washingtonpost.com/world/national-security/head-of-pentagons-secret-ufo-office-sought-to-make-evidence-public/2017/12/16/90bcb7cc-e2b2-11e7-8679-a9728984779c_story.html

Made in the USA
Monee, IL
23 February 2026

44799576R00149